THE HISTORY OF SCIENTIFIC IDEAS

A Teachers' Guide

THE HISTORY OF SCIENTIFIC IDEAS

A Teachers' Guide

Edited by DAVID STEELE, BSc, ARCS
Assistant Master, Winchester College

HUTCHINSON EDUCATIONAL

HUTCHINSON EDUCATIONAL LTD
178–202 Great Portland Street, London W1

London Melbourne Sydney Auckland
Johannesburg Cape Town

First published 1970

This book has been set in Imprint, printed in Great Britain
on Smooth Wove paper by Anchor Press, and
bound by Wm. Brendon, both of Tiptree, Essex

ISBN 0 09 089510 x (cased)
ISBN 0 09 089511 8 (paper)

Contents

SECTION III
SCIENTIFIC IDEAS IN BIOLOGY

DAVID HUGHES-EVANS, BSc, FLS, Farnborough Technical College

MAIN APPENDICES

Preface

There are many who, by their valuable criticism and encouragement, have contributed to this work. The authors would like to thank particularly: Mr Frank Beetham, Assistant Director of Education, Blackpool; Professor June Goodfield; Mr Rom Harré, Lecturer in the Philosophy of Science, Oxford; Professor F. R. Jevons, Professor of Liberal Studies in Science, Manchester; Sir Desmond Lee, Headmaster, Winchester College; and Mr T. F. Wheatley, Cambridge University Press. Also the members of staff of the following schools who offered practical advice and in many cases subjected their pupils to some of the material in this guide:

Bury Grammar School
Christ's Hospital
City of Bath Boys' School
Clifton College
Farnborough Technical College
Manchester Grammar School
New Mills Grammar School, Stockport
St Dunstan's College, Catford
Seaford College
Winchester College

We thank Mrs Shirley Hewitson for typing the final manuscript and the several other kind ladies who, often at very short notice, typed parts of earlier drafts.

In addition the Editor is indebted to Stephen Toulmin and June Goodfield for providing the opportunity to write this book; to Sir Desmond Lee for allowing the work to be done; to the Nuffield Foundation for the money to do it with; and to Mr John Stevens of Hutchinson for his unfailing help and patience throughout the enterprise.

The Editor welcomes criticism and comment on this guide. He would be particularly pleased to receive details of further books, articles, films and experiments which readers have found successful.

The College, Winchester, Hants DAVID STEELE

Acknowledgements

The following publishers kindly allowed us to use material from their books:

Princetown University Press and Harper Torchbooks for Table I.2 taken from *The Exact Sciences in Antiquity* by O. Neugebauer.
Macmillan for Tables II.1 and II.2 taken from *A Short History of Chemistry* by J. R. Partington.
Dover Publications for the extensive quotations from the preface to Lavoisier's *Elements of Chemistry* in the facsimile edition.
John Wiley for Table II.3 from *Inorganic Chemistry* by T. Moeller.
The *New Scientist* for Figures II.3 and II.4 from articles by Salem and Matthews in copies of 20 and 27 February, 1964.
In addition we should like to thank the publishers of the many books from which have been taken the large number of short extracts in this guide.

Introduction: The aims and use of this guide

Many people are being called upon to teach something about the way a scientist thinks and feels to people who cannot hope ever to acquire more than a smattering of scientific knowledge. Scientists and non-scientists seem hampered in their communication by a mutual incomprehension, and there is a sad lack of articulate scientists who can convey the excitement of the subject.

One of the best ways in which a layman in scientific terms can obtain some feeling for the way in which a scientist reacts is by considering the science of the past. This book offers an approach to science through some of the important ideas on which it is based. It is intended chiefly for those concerned with the teaching of science as part of general studies courses in schools and in further education. We hope it will help teachers already involved in such courses, that it will encourage others to start them and that it will serve as a useful 'primer' and source-book for anyone interested in the history and development of scientific ideas.

The concise text and the many and precise references will enable science teachers to make better use of the necessarily limited time which can be devoted to the preparation of 'general courses'. We hope also that arts teachers will be encouraged to teach those topics where the technicalities are not too great. The arts teacher can more easily gain the sympathy of non-scientists, and with scientists he may be better able to reveal new and stimulating questions to minds that are too often restrained by a limited and 'too tidy' formal training.

In classes made up of scientists and non-scientists the 'two sides' should be encouraged to help each other: the non-scientist can be made to articulate his problems and objections, the scientist to explain his subject in terms which the others can understand and appreciate.

It may be argued that by considering the problems of the past, problems which have been *solved*, science is bound to seem like a set of intellectual parcels, neatly wrapped up and finished with.

Certainly the study of past problems enables one to see patterns of discovery in better perspective, but it also affords an opportunity to relive the intellectual experiences—frustrations as well as triumphs—of our predecessors and to enjoy something of their excitement. We have tried to point out the blind paths, the faulty arguments, the prejudices of scientists and their times as well as the great steps forward in the development of ideas.

To appreciate fully these things will involve some discarding of modern notions and prejudices. This process is more difficult for scientists than for arts students and some may feel that to persuade the young scientist to put aside his hard-won ideas is an absurd enterprise. But very often the exercise leads to a much keener and more critical appreciation of modern concepts and the way they have developed.

We think it important that some attempt be made to cover twentieth-century ideas. The treatment must necessarily be superficial; but only by such attempts will both scientists and non-scientists be deterred from the opinion that science is an exclusively modern affair, all light and logic, owing nothing to the dark and superstitious gropings of the past. We hope that by discussing new ideas along with the old, students will sense not only the progress of science but also its continuity and will realise that the modern scientist is attempting to answer the same sort of questions as his predecessors, that his methods, though more sophisticated, are still uncertain and often fumbling and that his answers, though built on accumulated wisdom, are not final.

This guide is in the form of fairly short topics, each being designed to provide material for about four to six teaching periods. The material in each topic can easily be expanded to cover a longer session and the topics can be put together to form a course of any desired length and on any desired theme. (Some suggested schemes are put forward in Appendix A, p. 247). Though this treatment may seem fragmentary, the topics are linked within each section by the development of important ideas and generally by the recurring emphasis on matters such as intellectual environment, the balance between discipline and the imagination, the role of experiment and observation, the formulation of meaningful questions and the criteria of 'satisfactory' answers.

Only a limited amount can be included in each topic; we have aimed to provide something more than a skeletal outline but not copious teaching notes. We have been particularly concerned with those aspects whose treatment is not easily accessible to most teachers. When a subject is adequately dealt with elsewhere, only annotated references are given. It is important to realise this for otherwise it will seem that in several topics quite minor ideas receive equal or greater emphasis than more important matters: a guide is of greater use in unfamiliar terrain.

In appendices at the end of the book there are details of experiments and teaching aids, some suggested schemes and an extensive bibliography. The opportunities for practical work in the type of course suggested here are limited, yet to neglect experimental work is to lose sight of the essential nature of science. Whatever experiments are done it is most important that they are short (or can be done in short stages) and fairly simple (so that they can be clearly explained) and that they work! Demonstrations are valuable, for by actually making the same observations made by our predecessors we can more easily shed our prejudices and appreciate their interpretations of the facts.

In appendix F (p. 259) the bibliography to each Section is divided into three parts. The first contains those books which we feel the teacher ought to possess or have always available (N.B. page references to these in the text are to paperback editions where available); a good proportion of the books suggested in the second part should be consulted extensively—these might be put into the school library; in the third part are recommended books of more general and 'background' interest. There is also a general list of books and journals. Full bibliographical details are not given in the text; references consist of some easily recognised feature, generally the author's name. Where there are several books by the same author a shortened form of the title is given, e.g. *Architecture* refers to *The Architecture of Matter*. Each topic is numbered within its Section and internal references are in the form: section number.topic number (e.g. II.3).

This guide is the work of several authors and, while there is some uniformity in treatment and emphasis, it was felt undesirable to be identical in our approach. No attempt has been made to provide a comprehensive history of science; the limitations of space and the predilections of the authors have resulted in a *selection* of the more important ideas. Among the more obvious omissions are technology, the social sciences, medicine and mathematics. These could well form themes for other topics and we hope that our treatment will be useful as a pattern for these. Throughout the book philosophical issues are mentioned in their context but only briefly. Further possible points for discussion are suggested in appendix C (p. 252). We feel it would be dangerous to give too much detailed guidance on this, for precisely how much philosophical discussion occurs depends on the teacher and on the class. We hope our comments will form some useful starting points and that the books in the bibliography will help. But philosophical and sociological discussions must arise spontaneously and thus depend very much on 'class reaction'. It is hoped that a booklet will be made available for class use containing some of the quotations in the text, together with further exercises and quotations.

We intentionally have not included much biographical information (dates not in the text will be found under names in the index). This is something which the teacher can very easily find out for himself with the help of the bibliography. Biographical details and anecdotes can help considerably in giving interest and humour to the history of science, but there must be something more; we hope this guide provides it.

1 The nature of the universe

A. M. JOYCE and D. E. NEWBOLD

This section of the guide concentrates on the development of astronomy and mechanics. The subject commends itself because of its particular relation to the dichotomy between heaven and earth, and to the gradual movement towards a unified mechanical system in the Newtonian synthesis. A study of the development of astronomy to the present day can yield a striking picture of the struggles which man has had in assembling the mass of information he now holds; it provokes thought about the nature of reality and the 'truth' of the various planetary theories; it provides opportunities of investigating the social, literary and cultural environment of the times; above all, it provides a continuous picture of the development of a part of science for some 2,000 years.

We do not claim that our suggestions are much more than a foundation on which a teacher can build, but we hope they include a range of material sufficient for a coherent, comprehensible and convincing course in this branch of the history of science. We consider that the course should not be confined to 'chalk and talk' methods and from time to time we have indicated class experiments, demonstrations, films and discussion points. With the aid of these we hope the subject will be stimulating and provocative.

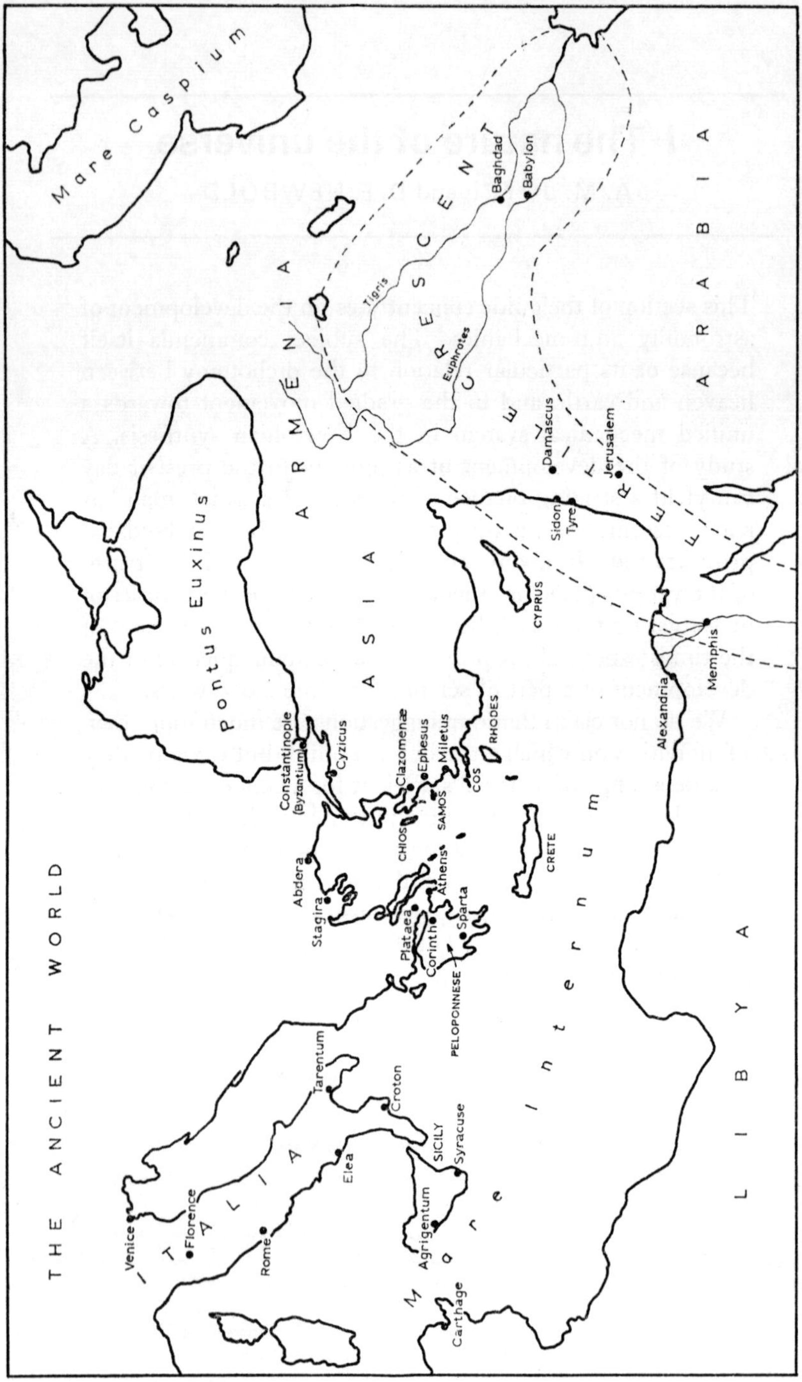
THE ANCIENT WORLD
Mare Caspium
Pontus Euxinus
ARMENIA
ASIA
FERTILE CRESCENT
ARABIA
LIBYA
Mare Internum
ITALIA
Tigris
Euphrates
Baghdad
Babylon
Damascus
Jerusalem
Sidon
Tyre
Memphis
Alexandria
CYPRUS
RHODES
COS
CRETE
SAMOS
CHIOS
Constantinople
(Byzantium)
Cyzicus
Clazomenae
Ephesus
Miletus
Abdera
Stagira
Athens
Plataea
Corinth
Sparta
PELOPONNESE
Tarentum
Croton
SICILY
Syracuse
Agrigentum
Elea
Venice
Florence
Rome
Carthage

Map I.1

I.1 Ancient astronomy

CHRONOLOGY

It is important when trying to understand the development of scientific ideas around the eastern Mediterranean to have some idea of the historical perspective of the times and the geography of the region. It is convenient, for instance, to speak of all the ideas of the people of Mesopotamia as 'Babylonian'—but this can lead to confusion. The earliest *mathematical* texts so far deciphered were in use in the time of Hammurabi (*c.* 1700 B.C.) when the Babylonian empire was at its height. The *astronomical* texts only reached their fullest development more than 1,000 years later, after successive empires had decayed and been replaced. By then writing and language had changed, and a Greek civilisation had established itself in the Mediterranean. Egypt too fell under the influence of different empires. Alexandria (founded 332 B.C.) rapidly became the centre of the Hellenistic world, and in particular of astronomical studies (see map I.1).

The principal chronological factors are shown in table I.2.

ASTRONOMY

A study of the astronomy of the Egyptians and Babylonians provides the opportunity to introduce many of the themes which recur throughout this course. Some of the following points should be thoroughly understood as the basis of all future work; others will probably only be touched on in discussion at this stage.

The astronomical observations

It is very important to make certain from the outset that everyone has a firm grasp of the astronomical observations which can be made with the naked eye, which successive theories sought to explain. These are:

(a) The daily apparent rotation of the fixed stars. (Many books, e.g. Berry, have long-exposure photographs to show this.)

(b) The annual apparent motion of the sun along the ecliptic against the background of the fixed stars—hence the zodiac.

(c) The apparent motion of the moon, explaining phases and eclipses, and the difficulty of reconciling lunar and solar calendars.

(d) The apparent motions of the planets, particularly their retrograde motion, variation in brightness, angular distances from the sun and periodic times.

These motions are probably best explained straight away in terms of a modern heliocentric picture, with the orbits of the planets in roughly the same plane, and with the plane of the moon's orbit inclined at approximately 5° to the plane of the ecliptic; but the motions on the celestial sphere must be thoroughly understood if the later arguments are to have any real meaning, and if the difficulty of reaching our modern understanding is to be appreciated. Tricker's *Paths of the Planets* deals admirably with all this and contains many valuable suggestions for practical work; Berry (ch. I) gives a good short summary in terms of the celestial sphere. A mathematical treatment (if required) is in Smart's *Foundations of Astronomy* (chs. V, X and XI): for convenience, the main astronomical terms used in the text are explained in the appendix (p. 89).

Measurement and recording

From a study of the astronomical data, one can go on to ask what events a regular observer would notice and think worth recording. These are *unusual occurrences* such as eclipses, occultations and conjunctions, and *regularly changing* ones such as the times of rising of particular stars, the motion of the planets, the motion and phases of the moon. One of the first requirements is a system for dividing the heavens into regions or constellations for easy reference. Both the Egyptians and the Babylonians did this and attached mythical and religious significance to them. Our constellations derive from the Greeks, who took many of theirs from the Babylonians.

The Egyptians recorded tables of star culminations and risings, and based their measurement of time on them. They also developed sundials and ingenious apparatus for measuring the azimuth of stars. But Egyptian mathematics, based on the 'unit fraction', was cumbersome (see the example on p. 22 and others in Sarton and Neugebauer). It impeded the analysis of observations and the prediction of future events. Early Egyptian astronomy seems to

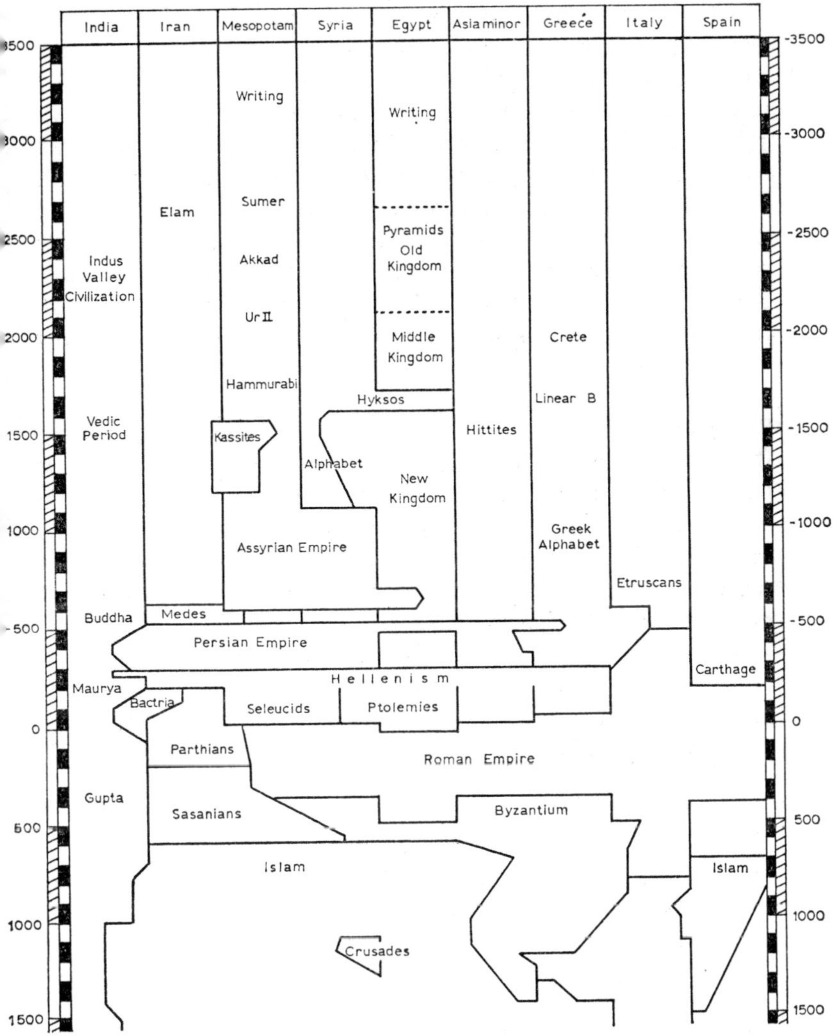
India
Iran
Mesopotam
Syria
Egypt
Asia minor
Greece
Italy
Spain
3500
3000
2500
2000
1500
1000
500
0
500
1000
1500
-3500
-3000
-2500
-2000
-1500
-1000
-500
0
500
1000
1500
Writing
Writing
Elam
Sumer
Pyramids
Old
Kingdom
Indus
Valley
Civilization
Akkad
Ur II
Middle
Kingdom
Crete
Hammurabi
Hyksos
Linear B
Vedic
Period
Kassites
Hittites
Alphabet
New
Kingdom
Greek
Alphabet
Assyrian Empire
Etruscans
Buddha
Medes
Persian Empire
Carthage
Hellenism
Maurya
Bactria
Seleucids
Ptolemies
Parthians
Roman Empire
Gupta
Sasanians
Byzantium
Islam
Islam
Crusades

TABLE 1.2

have been restricted to the marking of religious festivals and the measurement of the passage of time.

The Babylonians, however, had from at least 2000 B.C. developed a very successful mathematical system, with 60 as the basic unit (for details and an example, see p. 23). They gradually realised the importance of keeping careful records. From about 747 B.C. there are dated records of the movements of the heavenly bodies (used, for example, by Ptolemy in the second century A.D.). From about 450 B.C. they applied the sexagesimal system to angular measurement and divided the sky into twelve parts corresponding to the signs of the zodiac. From about 340 B.C. there are procedure texts which show how to use the long-term records to predict eclipses and future movements of the heavenly bodies.

Prediction

Because the movements of the heavenly bodies are *regular*, it is possible to predict future occurrences provided one's mathematical techniques are adequate. The Babylonians' techniques were only improved after Newton. Yet their cosmology, the explanation of what caused the observations, was extremely crude, even by contemporary Greek standards.

There is an illustration of the Babylonians' methods in the *Fabric of The Heavens* (p. 55), also greater detail in Neugebauer. For a general example of prediction of periodically recurring events, see the example based on the tide tables at the end of this topic (p. 24).

The written records

How do we know exactly what the Egyptians and Babylonians knew? Many references give good accounts of the discovery and decipherment of Mesopotamian clay tablets—a good subject for individual investigation. Sarton (ch. II) has a good, very brief account of hieroglyphics and the development of papyrus. One should make clear that our knowledge is partial—based on comparatively few Egyptian papyrus rolls and the decipherment of only a portion of the Babylonian clay tablets which have been discovered. When the contents of more of the latter are known, a fuller picture of Babylonian methods may emerge. There is good further background material in Olmstead.

The construction of a calendar

The problem of a lunar versus a solar calendar is discussed in detail on p. 86. The Egyptians very early adopted a solar calendar. Their year consisted of twelve months each of thirty days plus five additional days at the end, i.e. 365 days in all. The astronomical year began with the heliacal rising of Sirius, or Sothis (see appendix, p. 89), and hence consisted of $365\frac{1}{4}$ days. Thus after four years there was a difference of one day between the two calendars. After $365 \times 4 = 1{,}460$ years, the Sothic cycle was complete, and the two calendars were again in agreement. Knowledge of this permits the dating of some Egyptian events back to about 4200 B.C. This very logical system was adopted by Hellenistic astronomers, and was still used by Copernicus in his lunar and planetary tables.

The Babylonians, however, used a lunar calendar, some months having twenty-nine and some thirty days. To keep the lunar and solar years in step, they had occasionally to intercalate an additional month. Up to about 480 B.C. these intercalations show no regularity, but later the rule of seven intercalations in nineteen years at regular intervals was established. This is clearly a difficult system on which to trace the regularity of previously recorded events, but by no means impossible as long as complete calendar records are preserved.

Religion and astrology

Why should men bother to record the appearance of the heavens and theorise about them? Apart from pure speculation, the early astronomers were called upon to meet the needs of their primitive communities in attempting to predict natural disasters, regulate the seasonal nature of agriculture, and understand better the world in which they lived. The principal heavenly bodies were often thought of as deities, and religious festivals were associated with astronomical events, such as the new moon. Modern astrology will probably have been mentioned when discussing the zodiac. Its rise can be seen in many of the Babylonian tablets which contain messages predicting events such as the result of a battle from astronomical data. It is worth discussing how far the comparative separation of Greek and late Babylonian methods is due to the latter's preservation in exclusive priestly circles.

EGYPTIAN MATHEMATICS

(For a fuller account, see Neugebauer, ch. IV.)

The Egyptians used a decimal system with separate signs for unity and each power of ten up to one million, but there were no different signs for numbers between one and ten, and no zero. Hence there was no positional significance, and the various signs were written the number of times required to make up a large number. For example the number 3,421 would require 3+4+2+1=10 separate hieroglyphics. Though cumbersome, this system is quite satisfactory for keeping ordinary numerical records.

Calculations were simple extensions of ordinary counting. Multiplication by ten required the replacement of each sign by the one signifying ten times as much. Multiplication by other numbers was broken down into multiple additions. Division involved finding the number of times the divisor had to be multiplied to reach the original dividend. The best way of showing this is to carry out some of the simple calculations given in Neugebauer or Sarton by Egyptian and modern methods.

It is in their use of fractions that Egyptian methods appear most cumbersome. Apart from 2/3 and 3/4, the only fractions used by them were 'unit fractions' in which the numerator was 1. Other fractions when occurring had to be converted into the sum of a number of unit fractions. For example, the famous *Rhind Papyrus* begins with a conversion table of this type:

2/3	=	1/2	+	1/6	
2/5	=	1/3	+	1/15	
2/7	=	1/4	+	1/28	
2/9	=	1/6	+	1/18	etc.

The disadvantages of this method are that some fractons can be split up in more than one way, whilst others cannot be exactly expressed, except with great complexity. The Egyptians however seem to have acquired remarkable facility in handling them. Many of the problems solved in the existing papyri are the practical questions which must have occurred during the vast engineering works they undertook: volumes of earth to be moved, food and supplies to be divided between numbers of people, etc. Their methods continued to be widely taught into Hellenistic times. Though Ptolemy uses the highly preferable Babylonian methods in his calculations, he sometimes expresses his results in unit fraction form.

BABYLONIAN MATHEMATICS

(A very good account is in Sarton (ch. III). Neugebauer (ch. II) is more difficult.)

Early Babylonian texts use two cuneiform symbols, 𒁹 and 𒌋 to represent 1 (or a power of 60) and 10. Other numbers up to 60 are made up of combinations of these, e.g. 𒐊 = 5, 𒌍 = 30 etc. In the sexagesimal system, numbers above 60 make use of positional significance. Hence 𒁹𒌋 most simply represents $1 \times 60 + 10 = 70$. Until the late (Seleucid) period, however, there was no sign for zero, and hence no way of indicating which position represented which power of 60. 𒁹𒌋 could therefore also represent $1 \times 60^2 + 10 \times 60 = 4{,}200$ (the ambiguity could be removed by writing something like 𒁹𒌋o, using our symbol for zero). Similarly $1 \times 60^2 + 10$ ($= 3{,}610$) had to be written 𒁹 𒌋, i.e. with an empty space signifying the missing power of 60; and this too adds to the ambiguities.

The experienced scribe could no doubt overcome these difficulties. The large size of the base 60 makes confusion between one number and 60 or 3,600 times that number unlikely when taken in context. The great advantage of the Babylonian system was the way the positional notation was extended to fractions—the large number of factors of 60 (2, 3, 4, 5, 6, 10, 12, 15, 20, 30) making it particularly suitable for this. Thus our original example might also represent $1 + 10/60 = 70/60$ (or 70/3,600). Not all fractions can be represented in this way—1/7, 1/11, 1/13, for example, give recurring sexagesimal fractions. But the method is very powerful, and permits multiplication and division of large numbers to be done easily, once the multiplication tables for all numbers up to 59 are known. (Compare our decimal system, requiring the knowledge of multiplication tables up to 9×9—in fact if time and interest permit, this is a useful opportunity to discuss the whole question of counting on different scales. Modern applications with which pupils should be familiar are the use of the binary system in electronic computers, and the introduction of a fully decimal system of coinage and measures, with the resultant simplification of calculations.)

Using this system, the Babylonians developed a remarkable ability for solving problems, including the use of simultaneous linear equations with many unknowns, quadratic and some cubic equations. Tables of squares and cubes, square roots and cube roots up to very large numbers were computed, and areas and volumes calculated.

The other great achievement of the Babylonians was that having perfected a system of calculating based on 60, they then adopted a similar system of measurement, dividing the astronomical day into 360 equal parts, and the ecliptic similarly. Our division of the circle into 360°, each of 60′, each of 60″, derives from this. Thus they had a system for astronomical measurement and calculation comparable with a fully decimal system, but with a finer basic subdivision. As Sarton puts it: 'These gifts were perhaps too great to be appreciated by posterity except after an interval of many millennia.'

APPENDIX TO TOPIC I.1
THE USE OF TIDE TABLES

A useful example of cycles in nature can be obtained by a study of the tides. If a graph is drawn of tide height, day by day, for a year, for a place such as Avonmouth where there are large variations, the various periodicities will be clearly shown. Discussion can soon be provoked about why there are two high tides a day and about monthly and fortnightly cycles associated with lunar variations (Gamow's *Gravity* is useful here), and then about half-yearly variations associated with the position of the earth relative to the sun. The point here is that a study of actual happenings can lead very naturally to a close questioning of the reasons for these events. The value of the exercise can be increased if predictions are made of tide heights beyond the range of the graph and checked from appropriate tables. These examples of predictions of events and of enquiry into the reasons for the events can be compared with the roles of science as 'predicter' and as 'explainer'.

Examples of tide tables and of a graph drawn from them are shown below: this graph was drawn in one lesson by sixth-form pupils supplied with the appropriate data.

Tide data can be obtained from local tables in tidal areas or from *Whitaker's Almanack*.

Table I.3
Height (feet) of morning tides, Avonmouth, 1965

Date	*Jan.*	*Feb.*	*Mar.*	*Apr.*	*May*	*Jun.*	*Jul.*	*Aug.*	*Sep.*	*Oct.*	*Nov.*	*Dec.*
1	41.7	42.5	39.9	43.8	45.5	46.5	46.4	46.6	43.2	39.9	34.3	34.4
2	42.6	43.4	42.2	45.4	46.6	46.2	46.0	45.0	40.1	36.6	—	33.1
3	43.2	43.9	43.8	46.6	47.0	44.9	44.9	42.7	—	—	31.4	33.4
4	43.3	44.0	45.0	46.9	46.3	43.1	43.3	—	35.6	32.4	32.8	35.4
5	42.9	43.8	45.7	46.1	44.5	41.0	—	39.5	33.3	31.2	36.1	38.2
6	42.3	43.1	45.8	44.3	41.9	—	41.3	37.1	33.1	33.4	39.4	41.0
7	41.5	42.0	45.2	41.7	39.3	39.4	39.7	35.7	34.4	36.8	42.2	43.4
8	40.6	40.4	43.7	38.7	—	39.0	38.7	35.9	36.8	39.9	44.3	45.1
9	39.5	38.5	41.4	—	37.1	39.7	38.5	37.2	40.1	42.4	45.9	46.3
10	38.4	37.3	38.8	35.7	37.5	41.0	39.1	38.8	42.0	44.2	46.7	46.7
11	37.4	36.2	—	36.2	39.8	42.0	39.9	40.3	43.3	45.5	46.7	46.4
12	36.7	36.9	35.7	39.2	42.3	42.7	40.6	41.4	44.3	46.2	45.8	45.4
13	37.1	39.4	35.8	42.6	44.1	42.9	41.1	42.1	44.7	46.1	44.1	43.8
14	38.8	42.5	38.7	45.1	45.0	42.8	41.3	42.6	44.6	45.1	41.7	42.0
15	41.3	45.1	42.3	46.4	45.2	42.4	41.3	42.7	43.8	43.2	39.3	—
16	43.7	46.9	45.1	47.0	45.0	41.6	41.1	42.5	42.3	40.7	—	38.8
17	45.7	48.1	46.7	47.0	44.3	40.6	40.6	41.9	40.2	38.0	36.5	38.4
18	47.0	48.5	47.9	46.3	43.0	39.3	40.0	40.7	37.7	—	37.7	39.2
19	47.9	48.1	48.3	44.9	41.4	38.2	39.4	39.2	—	34.9	40.4	40.7
20	47.9	46.7	47.7	42.9	39.3	37.1	38.6	37.5	34.8	36.7	43.0	42.1
21	47.2	44.2	46.3	40.1	37.3	36.2	37.6	36.7	36.3	40.5	44.8	43.2
22	45.6	41.1	44.0	37.4	35.4	—	37.5	35.9	39.9	43.9	45.8	43.8
23	43.3	37.8	40.9	34.7	34.0	36.0	36.8	37.1	43.5	46.2	46.0	43.9
24	—	35.9	37.6	—	34.3	36.4	37.0	39.9	46.1	47.3	45.8	43.7
25	38.2	33.6	34.5	32.6	34.2	37.9	38.6	42.9	47.6	47.6	45.1	43.1
26	36.0	32.9	33.6	32.9	35.8	40.0	40.8	45.4	48.2	47.3	43.8	42.1
27	34.9	34.4	32.1	35.4	38.3	42.1	43.1	47.0	48.3	46.5	42.0	40.9
28	35.5	37.2	33.0	38.6	40.9	43.9	45.0	47.9	47.4	44.9	40.0	39.5
29	37.1	—	35.9	41.4	43.2	45.3	46.3	48.1	45.7	42.6	37.8	38.2
30	39.2	—	39.1	43.7	44.9	46.1	47.2	47.5	43.2	39.8	35.9	36.9
31	41.1	—	41.9	—	46.0	—	47.3	45.8	—	36.9	—	—

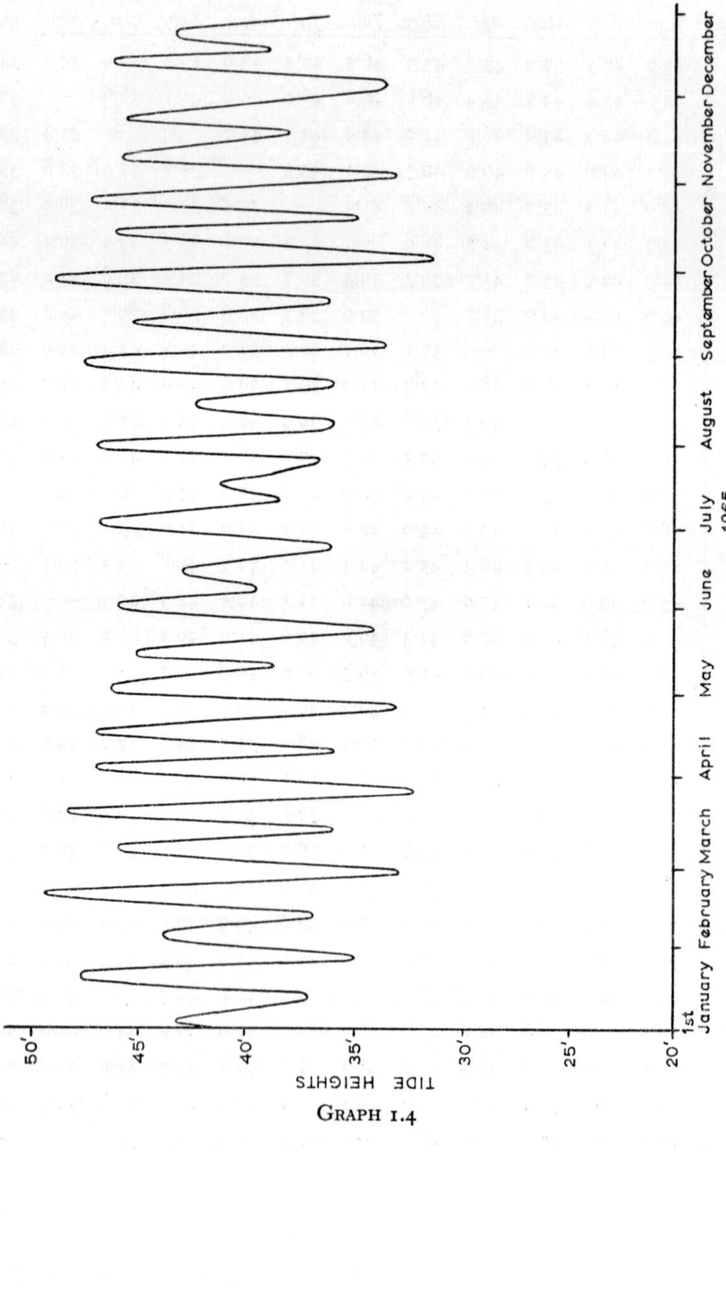

TIDE HEIGHTS
50′
45′
40′
35′
30′
25′
20′
1st January
February
March
April
May
June
July
1965
August
September
October
November
December

GRAPH 1.4

I. 2 Greek astronomy

THE EARLY THEORIES

What follows here should be read in conjunction with topic II.2 (p. 97). Even if it is not desired to spend too long on Greek science, the great significance of the first attempt to develop rational and general theories is best illustrated through the nature of matter as well as in cosmology and the laws of physical behaviour.

Again something should be said about the incomplete nature of our sources of knowledge. (*Fabric,* pp. 58–60; more detail in Dreyer, ch. I.) The classicists might be asked to contribute a short paper on the history of the recovery of classical texts.

The growth of scientific speculation

(*Fabric,* pp. 60–72.)

Abstract general theories about the nature of the world first appear in Greek philosophy, and one should try to discuss why. Was it the leisure and easier way of life, the geographical situation at the crossroads of empires and trade routes, the meeting place of peoples of many races, or the security from threats of invasion? The environment which gives free rein to speculation is not by itself enough. Why did Greek ideas persist and spread? The growth of open schools unhampered by ties to the temples is undoubtedly important. One should also examine the geographical distribution of the various Greek philosophers and trace the spread of their ideas (see map I.1, p. 16).

Early cosmologies

The earliest Greek thinkers whose cosmologies have been preserved at any length are quoted in the *Fabric* (pp. 72–7). They are:

Anaximander of Miletus (*c.* 610–545 B.C.), a younger contemporary of Thales (p. 73);

Anaximenes of Miletus, who flourished towards the end of Anaximander's life, during the Persian domination of Miletus (p. 75);

Anaxagoras of Clazomenae (*c.* 500–428 B.C.), the first Ionian teacher to move to Athens, where he became the friend of Pericles, later being banished for the impiety of his teaching (p. 76).

Compare these cosmologies with the earlier Creation myths. The attitude of comparing celestial and everyday phenomena was to prove very fruitful, leading to the first attempted measurements of cosmic dimensions. Students should be able to suggest the everyday observations on which these cosmologies are based. Notice also that each philosopher automatically thinks of the earth at the centre of his picture of the universe.

To give a fuller picture of the general intellectual flowering of Ionia at this time, it is worth mentioning also:

Xenophanes of Colophon (*c.* 570–470 B.C.), quoted as the first to discuss the significance of fossil remains in rocks;
Hecateus of Miletus (*c.* 550–475 B.C.), 'the father of geography';
Cadmus of Miletus (*fl.* 540 B.C.), one of the earliest historians;
Heraclitus of Ephesus (*fl.* 500 B.C.), the philosopher (see page 100).

Many fine works of engineering and architecture were completed in the same period, for example the 1,000-metre-long tunnel of Eupalinos on Samos, and the temple of Diana at Ephesus. Sarton, ch. VII (on Ionian science), presents an excellent picture of the period.

The Pythagoreans: the beginnings of mathematical astronomy

Pythagoras was born in Samos in the early part of the sixth century B.C. and was thus a contemporary of many of the Ionian philosophers. He is said to have travelled widely in Egypt and Mesopotamia before settling (about 530 B.C.) at Croton, a Greek colony in southern Italy, where he established a famous school, part academy, part religious brotherhood. He died in the neighbouring city of Metapontium *c.* 497 B.C.

The influence of the Pythagoreans in applying their mathematical discoveries to their theories of the nature of the world was profound and long-lasting. In particular note:

(a) the idea of *geometrical* perfection and harmony;
(b) the importance of arithmetical relationships; particularly those between whole numbers;
(c) the discovery of irrational numbers.

These ideas are discussed in the *Fabric*, pp. 77–87 (note especially Aristotle's account of Pythagorean astronomy, p. 80). A more detailed account is in Dreyer (ch. II). The whole significance of their ideas is dealt with very fully in Sambursky, ch. II, *Nature and Number*.

Myths and the development of rational thought

All the Greek philosophers agree that the world must be explained in terms of stable, eternal principles—however different the particular principles each one offers us. Yet this common idea was new.

The extreme varieties in the mythologies of their trading visitors may have led the seventh-century B.C. Ionians to look for some more rational and universal explanation of the world. The development of general mathematical theorems out of the old numerical methods may also have influenced them. One must not underrate the act of faith involved. In astronomy, the step does not perhaps appear so great. The regularly changing pattern of the heavens was well known in Egypt and Mesopotamia, and the early philosophers were great travellers. (The city of Naucratis in the Nile delta, originally a Milesian colony, was reorganised by Ahmose II between 569 and 525 B.C. and Greek trade encouraged. Ionia fell under Persian domination in the time of Cyrus (*c.* 546 B.C.). But to extend the idea to matter theory was far less obvious. On the earth's surface change, though continual, appears haphazard: it was not easy to imagine permanent principles of order lying behind it.

The contrast between Greek ideas and those of their predecessors is well brought out in Frankfort, particularly the first and last chapters.

The nature of the theories—topics for discussion

(See also topic II.2.)

(a) The Greek philosophers did not reach the stage of establishing or refuting their theories by appeal to *experiment* (though they were good *observers* of the natural world). This may make them seem very naïve, even stupid, to the modern student. One must remember that they were more concerned with putting forward a general picture of the world than in explaining particular events. There is the opportunity here to begin a discussion on the nature of scientific theory and what is meant by an 'explanation', also the difference between observation and experiment. Scientific students can be asked to contrast the methods of the Greeks with those of modern science. Their views on the latter are often in need of clarification.

(b) Very few actual experiments are in fact recorded by the Greek writers, yet an active technology existed alongside them. One sees here a divorce between scientific theory and technological practice which should, with its more modern history, make the subject of another discussion. (See also topics II.1, and II.2.)

PLATO, ARISTOTLE AND THEIR CONTEMPORARIES

Pythagorean ideas continued to develop in fifth-century Greece. Mathematical mysticism and the interest in harmony promoted a view of the world as a well-ordered and harmonious whole (the cosmos); the sphere, having perfect symmetry, was the perfect shape, the earth was round, and the universe was divided into the perfect superlunar and the imperfect sublunar regions.

Plato established his Academy at Athens in 387 B.C. His view of the purpose and methods of astronomy is shown in the passage from his dialogue the *Republic,* quoted in the *Fabric* (pp. 89–91). This brings out how he extended the legacy of Pythagorean ideas by his insistence on the importance of an ideal world, describable in terms of numbers and geometrical figures, lying behind the material world we observe. The stimulus which this emphasis gave to the development of mathematics and theoretical astronomy was immense. His pupils include Theaetetus, who proved that only five regular solids can exist, and Eudoxus, who devised the concentric sphere description of heavenly motions. But in belittling the role of observation and experiment, Plato's ideas exercised a powerful retarding influence on future development. His view that the very regularity of celestial behaviour indicated a divine nature reinforced the division between terrestrial and celestial phenomena which was only broken down again in the seventeenth century.

The concentric sphere theories

To Plato, the apparent motion of the fixed stars demonstrated the compliance of one part of the heavens with uniform circular motion.

His first attempt to explain the motions of all the heavenly bodies in this way occurs in the *Republic* (*Fabric,* pp. 92–4; see also appendix to Lee's translation of the *Republic*). The five known planets, the sun, moon and fixed stars are mounted on separate concentric rings, revolving at different speeds about a common axis, with the earth at their centre. This picture has two serious deficiencies: it fails to explain what causes the motions and regulates the different speeds (except in mythical terms) and, as Plato realised, it does not accurately describe the observed motions, failing particularly to give any retrograde motion for the planets and ignoring the obliquity of the ecliptic, which was also known at this time. Plato's ideas are further discussed in the introduction to Lee's translation of the *Timaeus*.

Plato's pupil, Eudoxus, succeeded in adapting the model to fit the observations better, but only at the expense of greater complexity. Eudoxus required twenty-seven spheres instead of the much simpler ring system: three each for the sun and the moon, four for each of the planets, and one for the fixed stars. Each heavenly body was considered to be mounted on the diameter of its innermost sphere, whose axis of rotation was fixed in the next outermost sphere which itself rotated at a different speed about a different axis, carrying the inner sphere with it—and so on for each successive sphere of one body. A full account of the system and a discussion of how far it actually fitted the observed motions is in Dreyer (ch. IV) and Heath (ch. XVI).

The following points should be emphasised in examining this theory:

(a) Any periodic motion, however complex, can be represented in this way, provided sufficient spheres are taken. Callippus, a pupil of Eudoxus, improved his master's system, which was seriously deficient in describing the motion of Mars and Venus, by adding one additional sphere for each (and also for Mercury). He also added two spheres for the sun and two for the moon, to allow for the different speeds at which they travel in different parts of their motions, making the total number up to thirty-four.

(b) The spheres of each heavenly body move quite independently of those of the others, without any mechanical links. Hence each has to have one sphere producing the daily rotation. A rotating earth would have reduced the number of spheres by eight!

(c) Eudoxus, like Plato, seems unconcerned with the possible physical nature of the spheres, their sizes, relative radii and composition, and what causes the rotation. His is a mathematical construction to describe the observations, not a physical theory. The relation of force and motion was confused at this time.

(d) With the work of Callippus and his successors in testing the theory of Eudoxus against the observations, and introducing modifications where necessary, scientific astronomy as we understand it may be said to have begun.

Aristotle

Much has been written about Aristotle's attempt to unify and record all the separate branches of natural philosophy, and of the tremendous influence his writings exerted until the rise of modern science. There is a brief outline at the beginning of the *Fabric* (ch. III). Before considering Aristotle's modification of Eudoxus' theory, one should examine his dynamical ideas. These are discussed in the *Fabric* (pp. 104–13), and more fully in Sambursky (ch. IV), 'The Cosmos of Aristotle', an excellent attempt to do justice to his views, which some writers, e.g. Berry and Dreyer, are inclined to dismiss very briefly. Aristotle was greatly influenced by his biological studies, in which he saw an orderly progression of growth and development to a final stage. This process, he argued, could only be fully understood in terms of the end to which it was directed. He extended this *teleological* approach to the cosmos, considering that the ultimate causes of the behaviour of all things could be found in the purposes for which they exist. In biology this method produced some useful results, but in physics it replaced the experimental investigation of the *immediate* causes of phenomena such as motion by involved speculation about the purposes for which they occurred. Modern science only progressed when it abandoned this approach. (See also topic II.2.

Examples of Aristotle's reasoning are:

(a) The primacy of circular motion.

Rotatory motion is prior to rectilinear motion because it is more simple and complete, which may be shown as follows: the straight line traversed in rectilinear motion cannot be infinite, for there is no such thing as an infinite straight line;[1] and even if there were, it would not be traversed by anything in motion: for the impossible does not happen, and it is impossible to traverse an infinite distance. On the other hand, rectilinear motion in a *finite* straight line is if it turns back a composite motion, in fact two motions, while if it does not turn back it is incomplete and perishable: and in the order of nature, of definition and of time alike the complete is prior to the incomplete and the imperishable to the perishable. Again, a motion which admits of being eternal is prior to one which does not. Now rotatory motion can be eternal, but no other motion can be so, since in all of them rest must occur, and with the occurrence of rest the motion has perished. . . . In rectilinear motion we have a definite starting point, finishing point and middle point. . . . On the other hand in circular motion there are no such definite points: for why should any one point on the line be a limit rather than any other? Any one point as much as any other is alike starting point, middle point and finishing point. (*Civics*, VIII, 9.)

(b) The necessity for a body's motion to its natural place in the cosmos.

On Aristotle's scale of cosmic values, the elements earth and water were confined to the imperfect sublunar region, while air and fire sought to reach the perfect superlunar region. The spherical shape of the earth was due to this tendency to conglomerate at the lowest (central) point in the cosmos. (*Fabric,* pp. 120–1).

(c) The relation between speed and effort. (*Fabric,* pp. 105–7).

(d) The relation between speed and resistance. (*Fabric,* pp. 107–9).

(e) The impossibility of a vacuum. (*Fabric,* pp. 109–10.)

Many of these questions can provoke lively discussions, provided they are presented fairly, showing Aristotle's development of the ideas of his predecessors. If they are appreciated at this stage, the difficulties of establishing modern dynamics will be more fully realised.

Aristotle's extension of Eudoxus' theory

(*Fabric,* pp. 114–20; Heath, ch. XVI.)

Aristotle applied his dynamics to Eudoxus' description of

[1] Aristotle argued that the universe must be finite in extent, otherwise the rotation of the extremities would require the traversing of an infinite distance in the finite time of revolution.

celestial motion, in an attempt to give a mechanical explanation for the movements of the concentric spheres. He required a source for the rotatory power, and this he took to be the natural tendency of the outermost sphere—the *primum mobile*. If the inner spheres were to derive their rotation from this, they had to be linked to it in a way that still permitted the movements described by Eudoxus. But linking the spheres of one body to those of the one further out would mean that the outer one would carry the inner one round with its own motion, unless extra spheres were included between them to counteract the movements of the outer body. For this purpose, Aristotle required one less reacting sphere than the number describing the motion of each body. (There being no other body within the moon's orbit, no extra spheres were needed for it.)

Thus the numbers of spheres required may be summarised as follows:

	Eudoxus	*Callippus*	*Aristotle*
Stars	1	1	1
Saturn	4	4	4+3
Jupiter	4	4	4+3
Mars	4	5	5+4
Mercury	4	5	5+4
Venus	4	5	5+4
Sun	3	5	5+4
Moon	3	5	5
	27	34	56

Criticisms of Aristotle's Theory

The main difficulties in this theory are:

(a) the increased complexity of the large number of spheres;

(b) the constant distances of each heavenly body from the earth fail to explain (i) the variations in brightness of the different planets (Mars and Venus are particularly noticeable), and (ii) the changes in the apparent sizes of the sun and moon.

These difficulties were realised in Aristotle's time, in fact in the theory of his contemporary Herakleides of Pontus (*c.* 388–315 B.C.) the ways of overcoming them were foreshadowed. Herakleides suggested:

(a) that the earth rotated daily on its axis;
(b) that Mercury and Venus revolved round the sun.

If applied to Aristotle's theory, (a) would have reduced the number of spheres required and (b) would have provided a varying distance between the earth and these two planets, although questioning the material nature of the spheres on which the sun was mounted, Herakleides' ideas were not supported by sufficient observational evidence, however, to be widely accepted, though they were revived and extended by later astronomers.

A far more revolutionary way of dealing with the problem was put forward by Aristarchus of Samos in the third century B.C. Aristarchus attempted to measure the relative sizes and distances of the sun, moon and earth geometrically. (The circumference of the earth was later measured by Eratosthenes—*Fabric*, p. 122.) Though his results were not accurate, they showed the sun to be far greater than the earth, and this may have prompted his suggestion that the sun was situated at the centre of the universe, while the earth rotated round it. Aristarchus also accepted Herakleides' idea that the earth rotated on its axis, but retained the idea of the moon's rotation round the earth. In order to account for the absence of stellar parallax, he suggested that the circumference of the earth's orbit was negligible in comparison with the distance of the fixed stars—a vast expansion of stellar distances. But though revolutionary and elegant, Aristarchus' ideas failed to explain the observations precisely enough—no simple system of circular motion could survive the rapid increase taking place in the accuracy of observations. The great astronomers Hipparchus and Ptolemy rejected his theory, and it was not revived before Copernicus.

For a general discussion of these ideas see *Fabric* (ch. IV), and Sambursky (ch. III). The detailed discussion and text of Aristarchus are in Heath. Dreyer (ch. VI) discusses Herakleides and Aristarchus.

HIPPARCHUS AND PTOLEMY

The development of later Greek astronomy is dominated by the work of three men: Apollonius of Perge (*fl.* 230 B.C.), Hipparchus (*fl.* 130 B.C. in Rhodes) and Ptolemy (*fl.* A.D. 140 in Alexandria).

Only the latter's major astronomical writings have survived. In their achievements one sees the progressive refinement of observations and the elaboration of mathematical theory which went with this.

Apollonius is credited by Ptolemy with the development of the theory of epicycles and eccentrics. The idea of the epicycle was contained in Herakleides' picture of an inferior planet rotating round the sun. As the sun moved round the earth, the planet would trace out a path whose speed and distance from the earth varied. Apollonius developed the geometry of this idea, and Hipparchus and Ptolemy used it to fit their observations. The eccentric is a circle whose centre is not at the centre of the earth. Apollonius showed that a suitable epicycle can be equivalent to an eccentric. This is fully discussed in Dreyer (ch. VII), and there are simple diagrams in the *Fabric* (p. 154).

A very good account of the great achievements of Hipparchus is in Berry (pp. 40–61), which explains the technical terms met in other works such as Dreyer (some of these are also dealt with in the appendix to Section I, p. 89). Hipparchus' work may be summarised as follows:

(a) the development of trigonometrical methods for astronomical calculations;

(b) the collection of a very extensive series of accurate observations;

(c) the use and critical comparison of a very long series of older observations, many Babylonian, to refine his measurements (see for example 'Lunar and solar calendars' and 'Precession of the equinoxes' in the appendix to Section I, pp. 86 and 87);

(d) the systematic application of the theory of epicycles and eccentrics to the observations of the sun and moon.

The accuracy he achieved is remarkable: for instance his estimate of the lunar month differs by less than one second from the accepted modern value.

Ptolemy continued Hipparchus' work, improving the accuracy of the observations still further. A good account of his work, and the contents of his famous treatise the *Almagest* is in Berry (pp. 62–73). More details are in Dreyer (ch. IX); and for those who are familiar with the astronomy, and wish to study the methods of the

Almagest in detail, Neugebauer has an appendix on the Ptolemaic system.

Two final points should be made:

(a) the system became increasingly mathematical, complex and 'unreal' as additional geometrical constructions had to be added to 'save the appearances';

(b) the Ptolemaic system dominated astronomy until the time of Copernicus, though the Aristotelian concentric sphere picture became incorporated in Christian theology and medieval thought. This fate of Greek learning during the Middle Ages is discussed in the *Fabric* (ch. VI), in Berry (ch. III), and in Dreyer (chs. X and XI), where the preservation of the Greek tradition in the Islamic empire during the Dark Ages is described. To the Arabs we owe also the introduction of *frequent* and regular observations of the principal heavenly bodies, the refinement of trigonometrical techniques, and many of our astronomical names: nadir, zenith, azimuth as well as zero, algebra and many others. Crombie's *Augustine to Galileo* gives a full account of the passage of Greek learning to the Arab world, and back into Europe from about the twelfth century onwards (see also topic II.3).

I. 3 Ptolemy and Copernicus

Although Ptolemy and Copernicus are separated in time by some 1,300 years, it may be profitable to consider together their work in the development of astronomical theory. Whilst in a purely factual account of planetary theory the Copernican system may be introduced as essentially a mathematical reorganisation of Ptolemaic theory, a student may well be encouraged at this stage to think of the more philosophic aspects of the transformation from the earth-centred to the sun-centred theory. In this topic emphasis is placed on the implications of and motivation for the new thinking.

PTOLEMY (A.D. *c.* 150)

Ptolemy's motives for the work he did should be emphasised. He was dissatisfied with the astronomical system of Aristotle, especially in its details. He was deeply concerned, as had been other Greek philosophers, to preserve the two fundamental axioms of Greek physics: circular motion and uniform motion.

On these two assumptions Ptolemy constructed models of the planetary system, using the epicycle, eccentrics and equants (his system is described in the *Fabric,* ch. v, 'Ptolemy's Mathematical Astronomy'). He was able to devise a very flexible system but not one which was of the same arrangement for each planet. He had to adopt different devices for each planet and abandoned attempts at a mechanical explanation of the resulting very elaborate and arbitrary system; in other words his theory explains *how* the planets move, but not *why*.

In the film *Earth and Sky,* a machine is used to show how the retrograde motion of a planet is an optical illusion resulting from our position as viewers on a moving earth. The machine is also used to show that the Ptolemaic and Copernican systems are geometrically and mathematically equivalent, so that from the results of simple *observations* alone no decision between them can be made. On a rather simpler scale, the idea of retrograde motion may be presented to students by considering the motion of a point on the extremity of the flange of a train wheel where, if viewed from a point outside the train, the movement of a point on the extremity of the flange will be backwards during a part of the revolution: this should be contrasted with the motion of the same point viewed from a position on the axle. It was the attempt to explain this retrograde motion that led Ptolemy to adopt his complicated system of epicycles and so on.

At this point, consider what evidence favoured firstly the Aristotelian model of interacting spheres and secondly the purely geometrical Ptolemaic model. Ptolemy justified his work thus: 'We believe it is the necessary purpose and aim of the mathematician to show forth all the appearances of the heavens as products of regular and circular motion,' and 'The circle hypothesis can be preserved and no one ought to regard the possibility of variations in the circle hypothesis and the introduction of a multiplicity of circles as unreasonable or illogical inasmuch as actual observations of the planet had disclosed irregularities which can be accounted for only by reasonable adjustment and adaptation.'

This question of the reasonableness or otherwise of the introduction of a multiplicity of circles is something which should be considered later in the light of the work of Copernicus. Mathematicians may like to compare Ptolemy's procedure with that of Fourier analysis. It is worth discussing whether the axioms of a scientific theory must be 'true', or whether it is sufficient if the deductions from the theory agree with observations. Although Ptolemy himself says that 'hypotheses are finally established and confirmed by the agreement of consequent proofs with the appearances', it may be argued that his own planetary theory is especially deficient in just this way. Given the position and velocity of a planet at a particular time, it is not possible to predict its future orbit with great accuracy from Ptolemy's theory: one can only account completely for the orbits already traversed by the planets.

The sanctity of the circular orbit, the abandoning of attempts at mechanical explanation of planetary motion, the importance of prediction in validating hypotheses: these are all aspects of Ptolemy's work which should be looked at carefully before passing on to study the Copernican approach.

COPERNICUS (1473–1543)

Copernicus too was not satisfied with the currently accepted explanations of planetary motion. He lived at a time of great intellectual revival when the work of the Greek philosophers was brought back for contemporary study. It may be that the publication of a translation of Ptolemy's *Almagest* about 1470, and also new astronomical observations about 1490 by Regiomontanus, caused Copernicus to attempt to reconcile the two and consequently to examine closely the whole of Ptolemy's theory. This resulted in his rearranging Ptolemy's planetary system so that the sun was at the centre of the universe. The epicyclic system was markedly simplified by reducing the number of components required to describe the motion of the planets: the number of circles required was reduced from about eighty to about fifty. This is a supreme example in science of the effect of looking at old facts in a new way, but its publication was long delayed by Copernicus, who seems to have feared public scorn for his fresh ideas. *On the Revolutions of the Celestial Bodies* was eventually published just after Copernicus died in 1543. The essay on Copernicus in Butterfield should be read at this point.

One may consider the motives for Copernicus' work by a study of extracts from it; these reveal the nature of his objections to the Ptolemaic system and his motives for trying to change it. Such a series of quotations follows, with phrases italicised (and numbered) to indicate that close attention should be paid to their exact meaning.

So I would not have it unknown to your Holiness that what moved me to consider another way of reckoning the motions of the heavenly spheres was nothing else than my realisation that Mathematicians were *not agreed* (1) about the matter of their research. For, first of all, they were so uncertain about the motion of the sun and moon and of the five other planets they did not in each case employ the same principles, assumptions and demonstrations of apparent revolutions and motions. For some used concentric circles only, other eccentrics and epicycles, by which nonetheless, they were not able fully to arrive at what they sought. For those who believed in concentric circles, although they demonstrated that several different motions could be compounded from them, could *not* derive from these motions anything certain that really *corresponded to the phenomena*. (2) Those who devised eccentric circles, though they seemed to have accounted for the visible motions in great part when their figures agreed with them; yet meanwhile they admitted that for the most part they seemed to *contravene the first principles*, (3) that of equality of motion [i.e. uniform velocity]. They could not find therein or deduce another most important thing, that is the form of the world and the certain symmetry of its parts [i.e. absolute circularity as an axiom].

. . . When therefore I had long considered this uncertainty of traditional mathematics, it began to weary me that no more definite explanation of the *movement of the world-machine* (4) established on our behalf by the best and most systematic builder of all, existed amongst philosophers who had studied so exactly in other respects the minutest details in regard to the sphere. Wherefore I took upon myself the task of re-reading the books of all the philosophers which I could obtain, to seek out whether anyone had ever conjectured that the motions of the spheres of the universe were other than they supposed who taught mathematics in the schools. And I found first, that according to Cicero, Nicetas had thought the earth was moved. Then later I discovered, according to Plutarch, that certain others had held the same opinion. . . .

The Pythagoreans, Heraclides and Ecphantus thought as much. . . . They conceived the earth to be turning at the centre of the world. . . . But if these views be accepted there arises another problem, no less important. What is the place of the earth? It is agreed by almost everyone that the earth is in the centre of the world. Yet if anyone were to deny this belief . . . if he should note that the motions of the sun and planets appear irregular because they are controlled with relation to another centre than that of the earth; he might perhaps be able to offer an explanation not superficially absurd *of the apparent irregularity* (5) in the motions of the heavenly bodies. . . . It would scarcely be surprising if someone were to attribute to the earth another motion besides the diurnal [i.e. daily] revolution. Indeed, Philolaus the Pythagorean, a remarkable mathematician, believed, they say, that the earth really moves circularly and at the same time executes other motions. He considered the earth itself merely one of the stars. . . .

When from this, therefore, I had conceived its possibility I myself also began to meditate upon the mobility of the earth. And although the opinion seemed absurd, yet because I know the *liberty had been accorded to others before me of imagining whatsoever circles they pleased to explain the phenomena* (6) of the stars, I thought I also might readily be allowed to experiment whether by supposing the earth to have some motion, stronger demonstrations than those of others could be found as to the revolution of the celestial sphere.

Thus, supposing these motions which I attribute to the earth later in this book, I found at length *by much and long observation*, that if the motions of the other planets were added to the rotation of the earth and calculated as for the revolution of that planet, not only the phenomena of the other followed from this, but also it is so *bound together* both the order and magnitude of all the planets and the spheres and the heaven itself, *that in no single part could one thing be altered without confusion among the other parts* (7) and in all the universe. Hence for this reason in the course of this work I have followed this system.

. . . Attacking a problem obviously difficult and most inexplicable, at length I had hit upon a solution whereby this could be reached by *fewer and much more convenient constructions* (8) than had been handed down of old.

In addition to considering the exact meaning of the portions *italicised* above, one should consider the following more general questions:

(a) What are the improvements since Ptolemy in attitudes towards the problems and general aims?

(b) What caused Copernicus to make these changes, bearing in mind that from the observational point of view there was nothing to choose between the systems?

In answering the questions and studying the quotations the philosophical as well as the technical factors should emerge. The passage numbered (2) lays stress on observations; passage (3) on the need to form a system in accordance with certain self-evident axioms; passage (5) defines the object of the theory and passage (6) insists that *physical* considerations should not interfere with the construction of axioms. It is noteworthy that in passage (6) Copernicus uses the word 'liberty' but then proceeds to confine his considerations to circular motions—why not elliptical also, for example? The answer, of course, is that Copernicus is still starting with the axiom of the Greek scientists (that of uniform circular motion), and he has to make his constructions and his account of the workings of things fit in with that axiom. By contrast, rather than start with self-evident axioms, contemporary science may be considered to frame hypotheses which *lead* to certain deductions, which are then examined to see if they are in accordance with the observable facts of nature. If we approach the study of planetary motion in this more modern way, the hypothesis of circular motion may be abandoned, but this was not done by Copernicus, and not until a century or so later, by Kepler.

Passages (7) and (8) mark the essential difference between Copernicus and Ptolemy, and these should be examined closely.

In (8) he stresses the need for his theory to consist of the fewest—that is, the most convenient or simplest—hypotheses necessary to produce accurate results. It must be remembered that both Copernicus and Ptolemy made the same assumptions: (a) circles and (b) uniform velocity. But the difference between them is that Ptolemy uses axioms of circular motion in such a way that he also has to bring into his theories a large number of supplementary circles for each special case considered. When Ptolemy has finished, he is not dealing with just a 'few and convenient' constructions but with a whole range of complicated, geometrical devices. On the heliocentric theory, developed by Copernicus, the whole situation appears very much simplified.

In his writing Copernicus claims also that his system contains a 'more rational system of circles'. It should be noted that this is a reference to the Pythagorean worship of number and form which were still important preoccupations in Copernicus' time.

Students should be clear about the extent of Copernicus' achievements (this is discussed in the *Fabric* (ch. VI) and in Butterfield). On purely practical grounds his system failed. It was neither more accurate nor in essence of greater simplicity than the Ptolemaic (see *The Watershed* for a full discussion of this). Questions which might be discussed are: 'Given that it was a failure, what made Copernicus insist on holding on to it?' 'Why did people immediately after him—such as Galileo and Kepler—become such confirmed Copernicans?' 'Was the real appeal of the sun-centred astronomy an aesthetic rather than a pragmatic appeal?' Aesthetic because (a) it could explain away the principal features of the planetary system without epicycles, and (b) it offered an explanation for the problem of retrograde motion. Pragmatically, (a) it explained the order of the planets from the sun, and (b) it gave an opportunity to estimate the relative dimensions of the planetary orbits in a way not possible under the Ptolemaic system.

The new approach to an explanation of planetary motion led once again to a debate between astronomer and cosmologist. Some insisted that Copernicus' theory was a mere technical device, not to be used except as a basis for calculation, and rejected the possibility that truth about the structure of the world could come from anything except divine revelation. This contrasts with

Copernicus' attitude that the truth about the heavens can be found by rational investigation.

One more thing should be emphasised: in the Copernican system, the fixed stars are now on an unmoving sphere. This leaves the question, 'by what forces are the planets moved?' The answer had to await the later contributions to inertial dynamics by Galileo and Newton.

Other topics which might usefully be discussed are:

(a) In what sense can Copernicus be regarded as the instigator of a revolution?

(b) In what sense was Copernicus looking backwards to the past, and in what sense was he looking forward into the future?

(c) In what sense was he tied by the observational and intellectual tools at his disposal? Was he fettered by the axioms of uniform perfect circular motion?

(d) What objections were there to Copernicus' ideas?

These may be considered in three ways: philosophical, religious and technical. A discussion of these points can be found in Hall, *The Scientific Revolution* (ch. 2).

Copernicus basically set out to solve the problem of motion of the planets, being dissatisfied with the predictive powers and with the complicated nature of the earlier Ptolemaic system. His single suggestion of a sun-centred universe made men look intently at the foundations of their world-picture: this may be regarded as his real achievement, even if it was not his primary aim.

I. 4 Tycho, Kepler and Galileo: astronomical advances

TYCHO BRAHE (1546–1601)

There are several things to notice about this Danish astronomer. Firstly, it was his ambition to re-map the sky accurately, and he was undoubtedly one of the greatest observers of all time, obtaining results within about four minutes of modern values. Photographs of his instruments—which were entirely for direction finding—can be obtained from the Science Museum. Secondly, his records emphasise particularly the continuity of his observations. Unlike his predecessors, Tycho observed every night, and in this way was able to build up a very comprehensive set of records. He tried very hard to detect stellar parallax, and it was his failure to do so that prevented his accepting the Copernican hypothesis.

His interest in stellar parallax arose in a particular way: in 1572, a new star was observed in the constellation Cassiopeia. Tycho, unwilling to lose hold of the astronomical traditions, was at first inclined to believe that the new star was in the sublunar region—the region beneath the moon's sphere—and therefore could be considered as a terrestrial object. Unlike objects in regions beyond the moon, it would thus be susceptible to change. Tycho argued that if the new star was in the sublunar region he would be able to detect parallax between it and the moon—against the background of the fixed stars. This he could not do, and so he concluded that the new star must be in the divine and unchanging heavens.

This left him with certain difficulties. If the star was in the region of the fixed stars, why did its brightness and colour vary? How could change in that region of the sky, which until then had been considered unchanging, be explained? A further problem arose later when Tycho tracked a comet against the background of the fixed stars, studying its position relative to them. He was forced to the conclusion that the comet moved in an orbit round the sun at a greater distance from it than Venus, so that the heavenly spheres could no longer be supposed fixed and solid, and another transitory body had appeared in the superlunary region.

His belief that the orbits of the planets could not be solid shells enabled him to advance his own theory of the planetary system. This was a compromise between the Ptolemaic and the Copernican, with the planets revolving around the sun, which itself moved in an orbit with the earth at its centre. Although he had seen the advantages of the Copernican system, his own observations hindered his complete acceptance of it. His objections included the absence of stellar parallax, the apparent distance of the stars and their differing brightnesses, and dynamical arguments against a moving earth.

Tycho's theory was quite widely accepted for a time, perhaps because, as he said, 'it would agree with both mathematics and physics, and avoid theological censure, and was at the same time wholly in accord with celestial appearances'. It was left to Kepler to put forward a new theory to explain planetary motion without recourse to the concept of solid spheres dating from Aristotle, but Kepler's ideas would have been impossible without Tycho's mass of data based on observation.

KEPLER (1571–1630)

The Sleepwalkers by Arthur Koestler is subtitled 'A history of man's changing vision of the universe'. This book is particularly valuable in a study of Kepler, especially as the part of the work concerned with him has been reprinted as a separate paperback volume entitled *The Watershed.* Parts of it are rather overwritten—for example: 'Lastly, they put an end to the nightmare that had haunted cosmology for the last two millennia: the obsession with spheres turning on spheres, freely floating in space, moved by physical forces acting upon them'—but in general *The Watershed* provides a useful basis for a study of Kepler. It is divided in this way: Kepler's life (chs. 1 and 3); the lives of Kepler and Tycho together (chs. 4, 5 and 7); relations between Kepler and Galileo (ch. 8); life story completed (chs. 9, 10 and 11).

Kepler's work can conveniently be divided into three stages. His first book, *Mysterium Cosmographicum,* is dealt with by Koestler in chapter 2. The most important stages in Kepler's intellectual life coincided with the book *Astronomia Nova* (ch. 6). This chapter is essential reading. Finally, chapter 9 discusses *Harmonice Mundi,* in which Kepler's third law is explained. Hanson's *Patterns of*

Discovery covers the discovery of Kepler's laws from a more philosophical, rather than psychological, point of view. This is well worth reading and comparing with Koestler's account.

When discussing Kepler and his work, there are three aspects to be kept clearly in mind. Firstly, there is Kepler, Pythagorean and mystic: convinced that harmonies exist in the world and that it is his job to discover them. Secondly, Kepler the mathematician: a man who made an intensive study of mathematics and geometry, and within these subjects made many discoveries, for instance an anticipation of calculus. Thirdly, and perhaps most important, Kepler the empiricist: passionately devoted to observations and to facts as he found them and, however mystical and mathematical, however much he cherished his imaginative theories, always ready to throw these overboard if they did not fit in with the observations.

The progress of Kepler's discoveries is clearly mapped out by Koestler and can easily be followed. Note his passion for harmonies, this Pythagorean–Platonic impulse which played a very important part throughout his life. His attitude towards Tycho's observations was also significant. Whatever he did, whatever theoretical orbits he had in mind, he always felt that these must fit in with the observations given to him by Tycho Brahe, for his master's accuracy was unquestioned. It was Tycho who gave him the observations of Mars, and asked him to work out the theoretical path of the planet.

Before he tackled the problem of Mars, even before he met Tycho, Kepler had made a number of interesting theoretical discoveries by asking a question which had not hitherto been asked: Why are there six planets, and why do they move with the periods they do? His answer was the first suggestion that a force emanating from the sun was responsible for planetary motion. He was trying to find out why the outer planets, which have a longer distance to travel to complete their orbit round the sun, travel at a slower rate along their paths. The problem of the mechanism involved was very much to the fore with the disappearance of the spheres as a result of Tycho's observations. Now, like Aristotle, he tried to assign physical causes to the phenomena and, as Koestler puts it, wedded physics and astronomy once more. There are two quotations here; one from the first edition of the *Mysterium*

Cosmographicum (1596) and the other from the second edition (1625).

If we want to get closer to the truth and establish some correspondence from the proportion (between the distances and velocities of the planets) then we must choose between these two assumptions: either the souls which move the planets are active the farther the planet is removed from the sun, or there exists only one moving soul in the centre of all the orbits, that is the sun, which drives the planets the more vigorously the closer the planet is, but whose force is quasi-exhausted when acting on the outer planet because of the long distance and the weakening of the force which it entails.

The following is taken from the second edition:

That such souls do not exist I have proved in my *Astronomia Nova.* If we substitute for the word 'soul' the word 'force' then we get just the principle which underlies my physics of the skies in the *Astronomia Nova.* For once, I firmly believed that the motive force of a planet is a soul. Yet if I reflected that this cause of motion diminishes in proportion to distance, just as the life of the sun diminishes in proportion to distance from the sun, I came to the conclusion that this force must be something substantial, 'substantial' not in the literal sense but in the same manner as we say that light is something substantial, meaning by this an unsubstantial entity emanating from a substantial body.

The changeover from the idea of souls which move the planets to the idea of a mechanical force acting externally upon them should be emphasised. Kepler himself thought this force was similar to magnetism, and when later he wanted to give an account of the forces involved, which would to some extent explain the elliptical shape of the orbits, he turned to a balance between the belief in a magnetic force radiating out from the sun, which rotated as the sun itself rotated, and the belief that each planet was itself a magnet whose axis always maintained the same line in relation to the sun (see Kuhn). As the planet revolved round the rotating sun the planet's two poles were alternately presented to it and, consequently, the sun itself being a huge magnet, the planet was alternately attracted and repelled. Being subjected to this alternate attraction and repulsion, the radius underwent fluctuations in length and this gave rise to the elliptical orbit.

Koestler is quite right to point out that men did not immediately throw off the old science and superstitions, but that there was gradual change, well shown in Kepler's case through the individual development of one single scientist. The *Astronomia Nova* is probably Kepler's most intellectually exciting book (see Koestler, ch. 6; *Fabric,* ch. VII).

In turning towards the new science, Kepler took up these questions, left unanswered by Copernicus:

(a) What is the exact shape of the orbit of a planet, and is this orbit centred on the sun, or is it eccentric?

(b) What mathematical formula describes the velocity of the planet?

(c) What is the relation between the velocity of the planet and its distance from the sun?

The three laws that answer these questions were not discovered in that order. The second law was discovered first, and the first law later, but both were published in *Astronomia Nova* in 1609. The third law was not discovered until 1618. The laws, put simply, are:

(a) Planets move round the sun, not in circles but in ellipses, the sun being at one focus.

(b) A planet moves not uniformly but in such a way that a line drawn from it to the sun sweeps out equal areas in equal times.

(c) The squares of the periods of revolution of the different planets round the sun are proportional to the cubes of their average distances from it.

Kepler's laws are by now called *laws,* not hypotheses or axioms. (Discussion point: What is the difference between these?) Basically, the difference lies in the fact that by now they are well established, having been confirmed by a very large number of observations. Secondly, and more important (perhaps this can be taken up again after studying Newton), they can be *derived* from a general physical theory—that is, from the Newtonian theory of gravitation. These laws are what one would expect from the theory of gravitation, but in order to arrive at a state of affairs in which he could formulate these laws, Kepler was obliged to discard a number of widely accepted axioms—not only those of the Aristotelian and Ptolemaic systems but even those of Copernicus himself. He threw out the axiom of uniform speed and as a result produced his second law of planetary motion. He spoke of a balance between the forces from the sun and from the planet, the sum total of which was to produce a sideways effect on the planet. He threw out the idea (from Copernicus) that the plane of the orbit of Mars oscillates in space, and was able to reduce Mars' orbit to terms involving only one circle, instead of the four of Copernicus. He almost came to the correct answer (or to what he felt was going to be the correct answer) when he could give the radius of an orbit with a permissible error of less than two degrees, and fit in all the correct positions of Mars for the ten oppositions recorded by Tycho. Then, at the end of five years' work and sixteen chapters, in which he leads one to believe that this is really his answer, he writes: 'Who would have thought it possible this hypothesis, which so closely agrees with the observed opposition, is nevertheless false?' And off he goes again for another three years' work. Finally, Kepler threw out the last axiom, that of circles, and started again; in *The Watershed* Koestler shows clearly how he may have come upon the ellipse.

Lastly, a student should consider briefly Kepler's last two works. The *Harmonice Mundi* illustrates his work on musical relations and harmonies, whilst in the *Somnium* he wrote about a journey to the moon. This is covered in Marjorie Nicolson's *Science and Imagination* (ch. 3, 'Kepler, the Somnium and John Donne'). Pupils will find it interesting to compare the details of the journey to the moon with both contemporary science fiction and contemporary space travel. It is also dealt with by Koestler (ch. 11, sect. 4).

Kepler is a good subject for careful study, both as a scientist and as a man. His passionate devotion both to his intellectual ideals and also to the observed facts provides an insight into all that is best in science.

GALILEO (1564–1642)

His work is best dealt with in two parts: his observational contributions, and his dynamical contributions which will be dealt with in a later chapter. In *The Watershed* the whole Kepler–Galileo episode is well told (ch. 8): the last years of Galileo's life are spoken of in *The Sleepwalkers* (pt. 5). These should be read, the last in conjunction with another book, *The Crime of Galileo,* by Santillana, an authority on Galileo. Brecht's play *Galileo* makes very good reading and might well be acted by a study group.

Examples of the Galilean telescope are in the Science Museum; photographs of these can be obtained. They can be constructed quite easily, and it would be a useful exercise to make some observations using a Tychonic direction-finder and a simple Galilean telescope. Although Galilean telescopes were used from about 1609 onwards, by the year 1612 the very necessary and good modification had been made of using a bi-convex lens as the eye-piece. By 1640 this was universal in telescopes; it was not until the second half of the century that sights were fitted. In 1610 when Galileo's book *Sidereus Nuncius* was published, he was still using the first simple Galilean telescope.

The effect of the discovery of the telescope on literature is something which might be discussed. *Science and Imagination* is a useful starting point. This book gives an assessment of the discovery of the telescope (a large number of countries claim this) as well as of the influence it had on the literature of the time. The preface is relevant and also chapters 'The Telescope and Imagination' (1), 'The New Astronomy and English Imagination' (2) and 'Milton and the Telecope' (4). If it is not possible for all students to read this, arts students may be prepared to give a lecture on the basis of these chapters. The poetry abounds with topics for discussion, and it is probably through the poetry rather than in any other way that one can experience a feeling for the immense assault on man's imagination and ideas at this time.

With regard to Galileo's observations, there is little difficulty. Firstly, and probably most important, the trouble about the apparent diameters was cleared up, even though no explanation was forthcoming. It was clear that what one saw through a telescope was no guide to the size of the object in question, the apparent size seen being determined by the instrument. This meant that the problem of the visibility of the stars was in effect shelved. Even though Galileo's observations broke up the sphere previously thought to hold the fixed stars—and granted even that the phases of Venus were observed, and that the planetary system of Jupiter seemed to provide a little model of the whole cosmos—nevertheless it was still quite possible for the critics of Galileo to insist that all these observations failed to prove that the earth moves. This was the central issue: Copernicus approached it from the point of view of mathematical simplicity, Galileo from a matter of physical observation. Taken alone or together, neither of these was yet enough to convince people. There were critics who refused to look through the telescope; and there were others who looked and refused to see anything. But there was also a third group which admitted the observations but insisted they did not prove the movement of the earth. They appealed back to Aristotle: only because he did not have good instruments at his disposal, they argued, was he unable to see the things Galileo had seen—but even if he had observed them his main idea would have been unchanged.

One of the points to remember and to emphasise is the contrast and conflict between those who believed that ultimate truth could be obtained through science, and those who believed it came only from divine revelation. It was really a conflict between these two extremes that led ultimately to the trial of Galileo, for in certain respects Galileo advocated the same kind of thing as Simplicius had many centuries before: that physics had no need of hypotheses, being concerned with things which in their very nature were true. This was something the theologians of that time could hardly be expected to agree with. St Thomas Aquinas says:

> The suppositions which astronomers have imagined are not to be accounted necessarily true. Although these hypotheses seem to save the appearances, we must not say that they are thereby proved to be facts, because perhaps it would be possible to explain the apparent movements of the stars by some other method which men have not yet excogitated.

This attitude is extremely sensible, but it was over this point that Galileo got into trouble. It seems that the Church did not mind him teaching the Copernican theory, provided he said the theory was merely a mathematical hypothesis. That he was loath to do so was possibly due to a number of reasons: firstly, he may have considered the Copernican theory to be true, in the way that any scientific theory can be said to be true; secondly, because he really did believe that there was nothing better to put in its place—no other hypothesis which fitted the facts better; and perhaps thirdly, because he did not wish to appear a party to the belief that study of the Bible and divine revelation was an equally good way of arriving at the truth. It may well be that the contrast between an hypothesis and ultimate truth was one he wanted to avoid. But underlying all this was the fundamental nature of the man—a forceful personality which led him to present a case perhaps more strongly than was justified by the evidence available to him: in this may lie the root of his disagreements with the Church.

The contrast between ultimate reality and what can be found out by scientific enquiry is not something that died away after Galileo and the Inquisition. Kant proclaimed it, and others, like Sir James Jeans, carried it on. Jeans, for example, declared that physics operates with pictures, that is actual or mathematical models, and because man can only operate with hypotheses, ultimate reality is hidden from us forever. Others, like Eddington and Poincaré, held similar views but for different reasons, and most of these theories divide the world into two: on the one hand, the laws of science by their very essence, by the way we construct them, force nature to take on a particular mode; on the other hand nature as it really is is something which exists but is forever hidden from us. There is plenty of scope here for class discussion; more comment can be found in topics II.2 and II.6 and in Stebbing's *Philosophy and the Physicists*.

By the time of Galileo, then, a state of affairs existed in which the Ptolemaic system was more or less defunct and the Tychonic compromise was generally accepted. With regard to the Copernican system, practically all mathematical astronomers were against it—with the notable exception of Kepler—and this may be an opportunity to ask the question: has it ever been *proved* that the earth *does* move round the sun? Is it not rather that a theoretical system based on a completely re-ordered dynamics has been created which at once is simpler and gives possibilities for accurate predictions, and which happens to include the movement of the earth? It is worth emphasising that in certain ways Galileo was conservative; in spite of the Tychonic observations, he denied that comets were heavenly bodies, and, perhaps even more surprisingly, he took little account of Kepler's planetary laws.

I. 5 The creation of mechanics

This section requires careful preparation if the main theme is to be made clear. There is not very much that can be done in the way of discussion, but practical demonstrations are possible, and can be used to bring out a number of important features. Here the principal changes are indicated together with topics suitable for practical work or detailed discussion. The *Fabric* (ch. VIII) is very useful for this section. *The Birth of a New Physics* should also be read and can be used as a class text.

EARLY DEVELOPMENTS

As mechanics and dynamics developed, ideas about *qualities,* inherent in Aristotle, were replaced by ideas about *quantities.* Aristotle had regarded qualities as fundamental, and indeed had regarded motion as a sort of quality. Gradually men began to replace these qualities by quantitative estimates and to deal with them quantitatively: this process is illustrated by the work of the scholars of Merton College, Oxford, in the fourteenth century (see *Fabric,* ch. VIII). As a first exercise students could list as many as possible of the qualities which can be treated graphically, and in this way quantified. For instance, the quality of warmth can be treated as that of a body possessing degrees of heat. One of the major tasks of natural philosophers has been this problem not only of trying to find quantitative solutions to problems of motion, but also of being able to formulate the problems in such a way that they *can* be solved quantitatively: this sums up the heart of the difficulties at this stage.

Notice the change from what might be termed 'essentialist' scientific theories to 'descriptive' scientific theories. A theory which described *how* changes in one quantity depended on changes in another came to be thought sufficient, without going further and asking *why* this was so. (Discussion point: Is a theory which only *describes* a particular phenomenon either satisfactory or sufficient? Laws of magnetic attraction and repulsion may be a useful example here.)

Motion comes to be regarded more as a state of a body and less as a process in which a body is concerned. (The Merton philosophers had this idea when they likened the continued motion of a rolling ball to the *gradual* cooling of a red-hot iron or to the *continued* sounding of a struck bell.)

Notice the confusion over the words such as 'force' and 'gravity'. Workers at this time operated with a variety of types of force, some of which are closely comparable with present-day forces, some quite different. 'Gravity' was used to explain free fall by supposing that there was a source of change *in the body*. According to them, there must be some kind of active cause which maintains the motion of free falling bodies; this they called gravity.

In these particular changes are the beginnings of idealisation: that is, the deduction of what the world is really like from certain unobservable theoretical ideals which have no counterpart in everyday life. Galileo's demonstration of a body uniformly accelerating down an inclined plane was in effect a demonstration of a theoretical idea applied to an everyday phenomenon. This experiment is described in the *Fabric* (ch. VIII).

One of the things to do practically is to repeat this and other Galileo demonstrations, one of which is illustrated in the film *Earth and Sky*. Here is the beginning of the idea of impetus, that property of the body which somehow *maintains* motion in an object; Galileo extracted this conclusion by imagining a world without any friction or air resistance. Students may devise demonstrations to illustrate the point that acceleration is the rate of change of speed with time, rather than the rate of change with distance. (Examples from elementary physics courses may be helpful here—especially *Nuffield Physics,* years 3 and 4.) Galileo's demonstrations took as an example a ball moving with uniform acceleration, in the way that uniform acceleration had been defined. It should be stressed that this was not a *proof*.

Scholastics, when studying free fall, were puzzled by the problems:

(a) Why does a body fall at all when nothing moves it?
(b) Why does it fall with increasing velocity?
(c) Why does it not fall with a velocity in proportion to the weight of the body, as it should according to Aristotle? (It is at this point that the legends about the Leaning Tower of Pisa come in.)

(d) Why does a stone which has been thrown preserve its motion after the initial propelling force has ceased?

THE CHANGE FROM CIRCULAR TO STRAIGHT LINE MOTION, AS AN IDEAL

There are two important phases in this change. There is the realisation that ideally bodies can keep moving forever without any external force upon them, but the question remains: in what kind of direction will they move? The reasons why Galileo believed that a horizontal direction was not a straight line but a circle around the earth are given in the *Fabric* (ch. VIII). The origin of straight line motion lay not in the work of the scholastics on mechanics and dynamics, but in the work of Descartes and the Corpuscularian philosophers, who utilised this for quite different philosophical reasons. An account of Descartes' system of the universe is to be found in the *Architecture of Matter* (ch. 7). The point to emphasise is that whatever went on in his universe did so in terms of a Euclidean framework. In the Euclidean framework the point of origin of anything is arbitrary, and there is therefore no reference to a centre. Descartes formulates a picture of particles endowed with motion by God in the beginning, having this motion which could continue indefinitely, with no reason, therefore, why a body once starting to move should ever stop. But, more important, he insists this motion is in a straight line, circular motion being produced only as a byproduct when one particle hits another.

It was clearly a revived interest in Euclidean geometry, together with a study of centrifugal forces (e.g. Benedetti studying the effect of a stone being held in a sling), which enabled men gradually to abandon the idea of natural motion always being in a circle and to accept the straight line instead. So gradually the emphasis changed again and the question became: what keeps things, which seem to move circularly, in their orbits? It was then only a short step to apply the problems of terrestrial motion as illustrated in this way to the problems of celestial motion. By the time of Borelli, there is a complete breakdown of the distinction between terrestrial and celestial laws. Both Borelli and Hooke realised that in order to explain the closed orbits of planets one had to suppose there was some force preventing them from flying off in a straight line out of the solar system. Borelli postulates not so much a force external to the planet, as the Aristotelian idea of the tendency of the planets to fall in towards the sun; but Hooke explains the closed orbit by supposing there is an attractive force within the sun constantly pulling the planets towards it.

Here a student should build up the pieces of the jigsaw which Newton was later to put together. The important thing to emphasise is that no one, not even Hooke, Huygens or Borelli, could demonstrate mathematically that their hypotheses would work. Admittedly, Hooke used a model to demonstrate that motions like those of the planets could be produced under the influence of a single central force, and this is a demonstration which would be very easy to reproduce. He performed it in 1666 at a lecture to the Royal Society. He took a heavy ball and suspended it from a wire in the ceiling, in such a way that the wire was free to move in any direction. When the pendulum was merely slightly displaced from its lower point, all it did was to oscillate backwards and forwards in a plane just like an ordinary pendulum. But when, after having been first displaced in one direction, it is given a sharp horizontal push in a direction perpendicular to this original direction, then it does not oscillate at all. What it does is to swing steadily in a horizontal plane executing a continuous orbit—and one can make the orbit either a horizontal circle or elongated like an ellipse by varying the strengths of the two pushes. The point of the demonstration is that the centrally-directed force, along the wire from the attachment to the ceiling, is unable to pull the pendulum back to the centre. All it can do is to deflect the motion towards the centre and thus produce a continuous curve. Given a similar force in the heavens, the effect on the movements of the planets should be the same.

In conclusion it is important to consider the changes in the physical problems of the planets since Copernicus. Basically they are these:

(a) the explanation of planetary motion has by now become a problem of applied mechanics, identical in principle with terrestrial problems of a similar nature, such as the pendulum and the projectile;

(b) experiments on the earth can be made to yield a direct knowledge of movements in the heavens;

(c) the distinction between the heavens and the earth, stated first by Aristotle and continued by Copernicus in *De Revolutionibus,* has started to break down. All the crystal spheres and the other devices have vanished, to be replaced by a terrestrial mechanism;

(d) more important, this mechanism shows that it can function quite as adequately as Aristotelian spheres.

Was anything of the medieval conception of the universe left? What problems remained to be answered?

I. 6 The Newtonian synthesis

Newton's role in the development of astronomy is well known; in one grand synthesis he brought together terrestrial mechanics and astronomical observation, applying the same natural laws to both, and ending the centuries-old dichotomy between the heavens and the earth. The way in which he developed this synthesis should be studied carefully, using one of the standard reference works for this period (e.g. Hall, *Galileo to Newton* and Cohen: Butterfield's account should be read also). An account of the development of ideas about gravity to modern times is given in Gamow's *Gravity*. The well-known story about Newton's discovery of gravity stands repetition here:

After dinner, the weather being warm, we went into the garden and drank tea, under the shade of some apple trees, only he and myself. Amidst other discourse, he told me, he was just in the same situation, as when formerly, the notion of gravitation came into his mind. It was occasion'd by the fall of an apple, as he sat in a contemplative mood. Why should that apple always descend perpendicularly to the ground, thought he to himself? Why should it not go sideways or upwards, but constantly to the earth's centre? Assuredly, the reason is, that the earth draws it. There must be a drawing power in matter: and the sum of the drawing power must be in the earth's centre, not in any side of the earth. Therefore does this apple fall perpendicularly, or towards the centre? If matter thus draws matter, it must be in proportion of its quantity. Therefore the apple draws the earth, as well as the earth draws the apple. That there is a power, like that we here call gravity, which extends itself thro the universe.

The germ of the idea of universal gravitation was held by Kepler, who had argued from the effect of the moon on the tides to the very likely greater effect produced in space by the larger earth. But he did not pursue this idea sufficiently far to link gravity quantitatively with astronomy.

By about 1685 Hooke too had a complete picture of the universe founded on a universal gravitational force: he had suggested this in *Micrographia* as early as 1665, and in 1678 he wrote: 'I suppose the gravitating power of the sun in the centre of this part of the

Heaven in which we are, hath an attractive power upon all the planets . . . and that those again have a respect answerable.' In a letter to Newton dated 1680, Hooke discussed the inverse square law in gravitational attraction.

Newton completed the *Principia* in 1686, after some considerable delay, partly occasioned by acute quarrels with Hooke who claimed credit for the discovery of the universal law of gravitation. In fact Newton had completed his *calculation* of the inverse square law by 1666. He checked his theory by comparing his result with observations of the moon's orbit around the earth, but at this time found a discrepancy which he later corrected using a revised measurement of the radius of the moon's orbit. It is the precision of mathematical analysis which distinguishes Newton's theories from Hooke's intuitions, and which has established Newton as the true author of the theory of gravitational attraction still acceptable today.

It should not be thought that Newton's work, even within the *Principia,* is confined to gravitational attractions. Through this work he exerted a lasting effect on scientific thought, not least in his definitions of mass and laws of motion, in his formal concepts of space and time, and in his establishment of theoretical physics as a mathematical science.

Besides Cohen's *The Birth of a New Physics,* the *Fabric* and *Architecture,* Hall's *From Galileo to Newton* is indispensable. The chapter on 'Newton and the World of the Law' gives not only a useful account of Newton's life, together with an excellent study of the way in which the *Principia* was written, but also full details of the controversies with Hooke and how this affected Newton himself. Newton himself deserves a fairly extensive treatment; for convenience, various aspects of Newton's life are arranged under headings which can be taken in whatever order is considered most suitable.

NEWTON THE MAN (1642–1727)

The life of Newton

Newton was born in Lincolnshire. He was a poor student who worked his way through college—he was sizar (a young man who waited at table) at Trinity College, Cambridge. During his studies he revealed himself to be such a brilliant mathematician that something unprecedented happened: his professor, Barrow, resigned in his favour when Newton was in his early twenties. Very early he published one or two books on problems of colour, and then nothing until the *Principia* at the age of forty-five. It is clear, however, that the elements of the intellectual solutions given in the *Principia* were in his head some twenty-four years before. He reworked the calculations in the *Principia,* publishing under pressure in about 1687, and seventeen years later he published the *Opticks.* He then became Warden of the Mint and its Master in 1699, called in as a working scientist to deal with the problem of the adulteration of metals. He held this post and that of President of the Royal Society until his death in 1727.

Controversies

Newton clearly disliked controversy and argument, though he was involved in both from very early on. In this dislike, many people have found the clue to his reluctance to publish. He quarrelled with Hooke on the problem of his interpretation of colour almost twenty years before the publication of the *Principia,* and also with Leibniz around the 1690s. Between 1704 and his death in 1727, several editions of the *Principia* and the *Opticks* were published, but it is clear that Newton was completely frank about his system of ideas only after the death of Leibniz in 1717.

Newton the introvert

Many books and speculative ideas have been circulating on this aspect of Newton. One which has had a considerable influence has been that of Lord Keynes, entitled *Newton the Man,* published in Cambridge in 1947 in connection with the Royal Society's Newton tercentenary celebrations. Another book—Manuel's *Isaac Newton, Historian*—reveals Newton's wider interest in such subjects as alchemy and theology, much more as characteristic of the breadth and outpouring of his genius than as aberrations of a sick man or a magician; it is a scholarly survey of his historical writings and his interest in the learned questions of ancient history. Newton can well be regarded as a man whose genius never allowed him to rest—a man with vast intellectual depth and an enquiring and very restless mind.

NEWTON'S INTELLECTUAL DEVELOPMENT

Newton is generally regarded as a mathematician, but in his time mathematics included not only the formal system we now understand by the name, but also the physical sciences, including matter theory. From his early days Newton shared with Boyle an interest and belief in a general natural system of corpuscles and forces, and he showed an early inclination to introduce the ether theory into this general system, to explain both gravitational and optical effects. He abandoned the ether theory in middle life for the following reasons:

(a) the problem of the interplanetary void: in this space between the planets there should be nothing to hinder their motion;
(b) embarrassments about the imponderability of the ether.

For an atomist it was preferable and simpler to equate matter with mass; if the ether had, in the case of the gravitation theory, been introduced to explain how two massive bodies interacted at a distance, it should itself not be a massive body, since it would then have gravitational influences. But if the ether was introduced to explain action at a distance and was itself matter, then it had to be matter without mass. Hence the problem of imponderability. The ether, however, was reintroduced somewhat agnostically in the later editions of the *Opticks,* to explain the mechanism of gravity.

NEWTON AND THE 'PRINCIPIA'

Forces

Newton set out in a systematic and mathematical way a system of ideas about dynamics and force which had been developing continuously since Galileo. The relevance of Descartes' early corpuscular system—that any deviation from straight line motion of a corpuscle implies that a force has acted on it—gives one the first law of motion. This Newton presented as a mathematical axiom.

Huygens had developed a theory of centrifugal force which implies Newton's second law of motion: that force is proportional to the rate of change of motion. The crucial idea for Newton was mass points and the forces between these points; it was the same idea that both Boscovich and Faraday were later to pick up (see topics II.6 and II.7). Newton applied this idea of forces between mass points to the probelm of gravitation with dramatic success, and then later, in the *Opticks,* extended it speculatively towards the problem of the matter theory. (See the *Architecture* and section II, p. 124, of this guide.)

THE ARGUMENT IN THE 'PRINCIPIA'

The essential problem was to establish an *analogy,* so that both terrestrial and celestial motion could be explained in gravitational terms. Newton was concerned to work out how gravitational forces vary with the distances between the attractive bodies, and how a knowledge of this attraction could be used to account for both terrestrial and celestial motions.

His argument is conducted in two or three steps:

(a) He begins from a study of the projectile and the forces which cause a stone to fall to the earth.

(b) He extends this to a consideration of what men knew at that time about planetary movements, utilising Kepler's laws and explaining them on the hypothesis of a single centrally-directed force which varies inversely as the square of the distance.

(c) He then uses mathematics extensively to explain how fast a planet must fall towards the sun, or the moon towards the earth, in order to remain in a stable orbit. This involves complicated calculations and the relation between the masses of the heavenly bodies, their velocities and their distances from the earth. It would be constructive for students who are especially interested in space rocketry to try to find out just what calculations of masses, velocities and attractive forces are required in order to launch a satellite successfully.

NEWTON AND HYPOTHESIS

Newton starts with certain initial assumptions, which at the end may be more or less weakened. His method of hypothesis is almost a casebook illustration of a certain type of argument involving a formal fallacy; it would be very useful for students to be faced with this difficulty and to explain it. We say that if Gravity G is assumed, then certain phenomena $A1$, $A2$, $A3$ should follow. But $A1$, $A2$, $A3$ are in fact the case, so presumably gravity G is authentic.

This is very like the formal fallacy in the following: If Professor

Jones is a Communist he will be against lynching; but Professor Jones is against lynching, so presumably he is a Communist. Explain the difference between these two arguments which, on the face of it, look extremely similar. They are both arguments from analogy, but (a) there are many phenomena *A*1, *A*2, *A*3 which are explained by gravity. There are many *A*'s but only one *G*. In the case of Professor Jones there is only one *A*—one phenomenon, that is against lynching—but there are many alternative *G*'s which need to be eliminated before one can arrive at the true explanation of anti-lynching. And (b), whereas, in the case of Professor Jones, there could be a number of equally powerful alternative explanations as to why he is against lynching, in the case of the phenomena *A*1, *A*2, *A*3 there is no *powerful* alternative explanation other than gravity.

One further point about the character of this hypothesis and the distinction between the kind of laws Newton formulated and those of Kepler. Kepler's laws are factual, categorical discoveries; Newton's laws are far more concerned with hypothetical relations. Newton carried to the extreme the ideas of mathematical idealisation—starting with abstract ideas with no application in real life and in the theory of motion, carrying this to a far greater limit than earlier workers in this field had ever done. By systematically applying these ideas he was able to bring together the work of his predecessors in one tremendous synthesis.

Consider the nature of this type of man, with a far-ranging mind, who can bring together an enormous body of work—some of which has hitherto been quite unrelated—and from it build up a comprehensive picture. Students may compare Newton with other scientists who did the same thing, or make analogies with men in other artistic fields, with the same kind of attitude. In Duhem's book *The Aim and Structure of Physical Theory* there is a section in which he contrasts the English and French attitudes towards physics in the nineteenth century and compares this difference with the differences between Shakespeare and Racine. Is there a national character in the intellectual field which reveals itself both in the structure of scientific theories and in the artistic field in the nature of the artistic works produced? This discussion can be carried over into the realm of music and painting as well as science.

NEWTON'S ACHIEVEMENT

Halley's poem is quoted in the *Fabric* (ch. IX) and is worth discussing. Students may like to explain this, discussing in particular the last line, 'Nearer the gods no mortal may approach'. Would they feel that of all human activity, scientific activity as exemplified by Newton would bring men nearest to an understanding of the Creation, or, as an alternative interpretation, put men in the position of God?

The relationship between practical astronomy and Newtonian theory should be brought out. The theory makes the observations intelligible because it makes sense of the patterns the practical techniques have revealed. But it does not necessarily improve the techniques, it only explains their success—and this is certainly a great merit of the theory. It does not matter that one does not then rework the techniques on the basis of the new theory. If any mathematics students are energetic enough to attempt to calculate the tide tables on the basis of Newtonian gravitation, let them try to do so!

After this, students could compile a table: comparing and contrasting, in as much detail as possible, the geocentric medieval system of the world and the completed heliocentric picture as now formulated by Newton. Historians could prepare a paper on the origins of the Royal Society and the part that societies and academies were playing in the development of science at that time. The standard work on this subject is G. N. Clark's *Science and Social Welfare in the Age of Newton*. As an extra exercise, mathematicians might try to produce a formula relating the period of a satellite to its distance from the earth and, using this formula, check the estimated distance of the earth from the moon. At this point, a discussion on a number of questions which arose immediately out of the gravitational theory and its success may be appropriate.

Theological difficulties: begin by discussing the question which was immediately posed to Newton by Leibniz, who found his system of philosophy totally unacceptable. Basically, the argument was over the question of interference by the Creator. The two problems about the Newtonian theory raised by Leibniz were:

(a) Newton's system was certainly comprehensive, but it did not explain how the planets got into their orbits in the first place;

(b) the cumulative effect of the slight perturbations in the planets should be sufficient over a long period of time to cause the planets eventually to go completely off course.

Now a question arises over the role of Providence in nature. Does God show His power and wisdom by intervention in natural powers, or by arranging things in the first place so that He never needs to intervene? The basic trouble with the Newtonian philosophy seemed to be that, according to Newton, God had provided a rather botched job in which He needed to intervene. The question to discuss is: Can one have it both ways? Try to interpret the difficulties left over by the gravitational hypothesis (*Fabric,* ch. x). Newton did try to have it both ways; he was not theologian enough to face directly the question of intervention.

NEWTON AND THE THREE BODY PROBLEM

This sometimes proves puzzling. There is no general formula which can express the results of three bodies acting simultaneously upon each other. Moreover, there is a proof—but a modern proof and not contemporary with Newton—that there can be no general formula. One can calculate numerically a solution for any three given bodies, if their masses, velocities and distances are known, but one cannot find a general expression for this problem. This, too, was an obstacle to the reception of the system.

NEWTON AND THE PLANETARY INEQUALITIES

The story of the discovery of Neptune and the non-discovery of Vulcan is worth investigation (see topic I.7, p. 76). It is an excellent example of how a theory may have to be modified. If the theory is true, then something (in this case, a planet) which is causing the aberration should, on prediction, exist. If the aberrations are there and the predicted phenomena cannot be found, then one can only conclude that the theory does not hold good all the time—and this leads directly to Einstein and the ideas of relativity.

NEWTON AND THE MECHANISM OF GRAVITY

It is well worth pointing out once again the nature of this problem. Remember how often the idea of a 'mechanism' of some kind has been discussed in this history. Wherever questions of action at a distance come up, questions of mechanism immediately arise. It is always natural for physicists to ponder and ask these questions of mechanism, and the whole form of the argument over the mechanism of gravity which took place in the eighteenth century arose again in the nineteenth over problems of radiation (see II.7).

It is sometimes easier to appreciate the difficulties involved if science students are set against arts students on this point, and asked to try to explain how gravity works. The arts students should be as difficult as possible over this question. Or ask the scientists to explain how light travels, or how magnetic influences travel, or why, if sound waves need a material for their propagation, gravity needs no material, or why electromagnetic radiation appears to need no material. The whole problem of action at a distance ranges not only across the field of magnetism or gravity, but also into the realm of matter theory and the problem of cohesion and forces between atoms (see Hesse for a full account). This is very relevant to the problems inherent in a pure atomic theory, which may already have been stressed in the relevant work on Greek science.

Lastly, an examination of the fundamental difficulties of the Newtonian theory can bring us into a relativistic position (*Fabric*, ch. x). Especially significant for an understanding of the ultimate difference between a truly heliocentric relativistic picture of the world and a geocentric one is the realisation that distinctions of absolute space and time in centre cannot be drawn sharply, and that on a more refined picture of the universe all massive bodies are centres of gravitation relative to one another, rather than any one of them being *the* centre of the universe.

THE WIDER IMPLICATIONS OF NEWTON

Encourage pupils to prepare a paper on the reception of Newton's ideas and subsequent intellectual developments in Newton's age. For this, Hall's *From Galileo to Newton* (ch. 11, 'The Age of Newton') is excellent. In this chapter Hall considers the Leibniz-Clarke correspondence and also shows very clearly why acceptance of Newtonian attitudes in France was so long delayed. Basically it raises again the issue of mechanism. In the whole of Descartes' system there was no room for action at a distance: changes of direction occurred by impact. A second objection—equally influential—was that, whereas Newton's mathematics were undoubtedly unassailable, as a contribution towards the physics of the heavens they were inadequate because one could not tie up any physical reality with these mathematical theorems. Questions which may be considered are these: In a philosophy of nature, is it enough to have defined mathematically the action of natural forces in terms of the effects they produce? Should one go further and try to explain what the *causes* of those natural forces are? Leibniz was not prepared to accept mathematical equations as the ultimate truth in human knowledge; Newton was afraid of doing so. Hall, in *From Galileo to Newton*, shows the seventeenth-century dilemma: on Cartesian principles the universe is mechanistic; if the universe is wholly mechanical, then God has no part in it; if God controls the universe, then it is not mechanical, nor is it governed by laws. How, if at all, did Newton, Leibniz, Descartes and the seventeenth-century philosophers reconcile this position? What are the implications of this dilemma in twentieth-century science? It does no harm to promote a theological discussion here. Is one forced into the position of supposing that the universe is either mechanical or miraculous? Leibniz would say that Newton had lessened the power of God and undermined religion, and also that he had at the same time a subversive and mechanical philosophy—that is, the philosophy of Descartes.

NEWTON AND LITERATURE

Jonathan Swift's well-known *Voyage to Laputa* is a satire on the Royal Society and contemporary scientific problems of the time. The relationships can be found in Marjorie Nicholson's *Science and Imagination* (the chapter on the scientific background of Swift's *Voyage to Laputa*).

NEWTON AND ASTROLOGY

Kepler was about the last of the great scientists to earn a living by astrology as well as by science. It may be said that astrological belief is not a result of underestimating physical influences at work in the world, but in fact of overestimating them. Pre-scientific people may have thought that everything in the universe was capable of acting on everything: what Newton did was to sort out which of those effects were authentic. In contemporary science we insist that the authenticity of claims relating natural events at one point with natural events at another should be demonstrated and underlined. Until some convincing demonstration is forthcoming, we tend to be agnostic about such claims. A relevant and interesting example is the attitude of most scientists towards extra-sensory perception, and the demonstrations in *Of E.S.P. and Telepathy* by J. B. Rhine. Students may like to devise a list of events which could possibly be related physically in some way, and to devise tests to demonstrate that these events are related. Telepathy is a very good example: it gives an opportunity to see just how difficult it is to devise crucial experiments in which the likely causal factors are isolated from each other. (For Newton and alchemy, see topic II.4, p. 124.)

THE PROBLEM OF ETERNAL PRINCIPLES

The last section of the *Fabric* deals with questions which may have been posed very early on when considering Greek science. With specific reference to Newton, the problem is: how far can a physical theory be mathematically certain and coherent and at the same time be descriptive of the world as we find it? The problem was stated slightly differently in two different centuries: first by Vico, arguing that mathematical systems do have certainty and coherence just because they are our own intellectual creation rather than something outside ourselves to discover; second, by Einstein, who wanted to argue that there is a certainty about mathematical axioms which may be absent in the theoretical principles we formulate. This question was: If our theoretical principles are statements of how we find the world to be, can they at the same time have the certainty belonging to mathematics, belong to such theorems as we find in Euclidean geometry, with their logical dependence one upon the other?

These points and problems have formed a fairly continuous thread throughout the development of astronomy. It does no harm to end this section on a philosophical note, because from the study of the developing concepts of the universe one can deal with fairly philosophical questions at a level suitable for many students. If possible, they should all read Toulmin's *Philosophy of Science*; if any moral is to be drawn, one would hope that the pupils would realise, perhaps for the first time, how important in the growth of science has been the re-ordering of intellectual systems and the formulation of concepts. As an excellent way of completing this section of the course, students may discuss the question: in the development of our ideas about the universe, what have been the various roles of observation, experiment and the formation of intellectual concepts?

I. 7 The modern universe

This part of the course might be taken separately, or as the starting point for a series of backward excursions in the history of science, or, as here, as the culmination of a historical development. Our modern knowledge is so remarkable, and the theories based on it so exciting, that the development of observation and reasoning on which they rest should be clearly understood by everybody. One can bring out the remarkable progress made in the last century, and particularly in the last fifty years, by reminding students:

the info was gained by

(a) that our information is based solely on the radiation received by the earth from these immensely distant objects, and by applying the terrestrial laws of physics;

(b) that prior to 1838, the distance of even the nearest star had defied measurement;

(c) that the chemical constitution of the stars was considered, at least as late as 1825, as an indisputable example of knowledge that would forever remain unknown to mankind (e.g. by the French philosopher, Auguste Comte).

In contrast, a brief 'tour of the universe' at this stage, based on modern information, should impress even the most blasé student. We now know not only the distances of the very distant nebulae, but also the chemical constitution, size, mass, density, temperature and energy radiation of many stars, and can postulate the processes occurring within them as they go through their life cycles. The light year must be used as the unit of measurement for the vast distances concerned, and some feeling for the time scale of cosmic evolution should follow from this. Notes on a more detailed discussion of the age of the universe follow later in this topic.

The information for all this is readily available in many modern books, such as Lyttleton's *The Modern Universe*, Hoyle's *Astronomy* or Larousse, and the facts can be copiously illustrated with photographs from these references, projected either by epidiascope or overhead projector, or from slides.

The steps in the building up of this knowledge can now be dealt with in more detail. Here they are summarised as follows:

the development of observational techniques;
the measurement of distances by stellar parallax;
the application of gravitational theory;
spectroscopic analysis;
temperature and density measurement;
types of stars;
energy mechanisms;
further distance measurements;
the distant nebulae.

For details, two outstandingly useful references are Vaucouleurs, *Discovery of the Universe,* and Smart, *Some Famous Stars.*

From this more detailed knowledge, it is possible to go on to discuss theories of the age and evolution of the universe, the formation of our planetary system, the conditions necessary for life on other planets, and the possibility of life in other parts of the universe. In conclusion, something might be said about the theory of relativity, though this is a difficult topic, and about some of the implications of space travel.

THE DEVELOPMENT OF OBSERVATIONAL TECHNIQUES

Telescopes

The best reference here is King, *History of the Telescope,* which describes and assesses the principal telescopes, with numerous photographs and a wealth of detail about the astronomers as well as their instruments. For quick reference to the astronomical findings, Vaucouleurs and Berry (who ends with the nineteenth century however) are very useful. There are numerous good pictures in Hoyle's *Astronomy* and in Larousse.

The more accurate measurement of stellar positions is not the principal reason for the development of bigger and better telescopes. This depends on the accuracy of the mountings and the method of dividing the angular scales: Tycho Brahe, using simple sighting devices, achieved an accuracy of about one-half minute of arc (his apparatus is described in King). Nor is magnifying power the most important consideration, except for the detailed study of the solar system and of double stars, since the great distances of all stars make them effective point objects anyway. The two most important characteristics of an astronomical telescope are high resolving power, i.e. the ability to see as separate two close objects, and light gathering power—the ability to form a recognisable image of a very faint object. Both these requirements are best served by a telescope of very large aperture, and the technical problems of achieving this are the history of telescope development. The other requirement is favourable atmospheric conditions—hence the siting of the large American reflectors in the Californian mountains, and the comparative uselessness of the Earl of Rosse's large reflector in Ireland. There is plenty of scope for demonstrations here—of simple telescope types, variations in resolving power with aperture, as well as actual telescope observation of the moon, planets, the Andromeda nebula and some of the better-known stars.

Galileo's telescope had lenses of only a few centimetres diameter, but in comparison with the pupil of the unaided eye this represented a big improvement—shown by his immediate observation of Jupiter's moons, the starry nature of the Milky Way and the many other stars invisible to the naked eye, as well as the mountainous appearance of the moon, the phases of Venus and the peculiar appearance of Saturn. The normal astronomical telescope with a converging eyepiece soon came into use, and compound eyepieces increased the magnification obtainable. To reduce aberrations, very long focus objective lenses were introduced, resulting in lengthy and unwieldy telescopes. The recognition of the importance of telescope observations is shown by the establishment of the Paris Observatory in 1671 and the Royal Observatory, Greenwich, in 1675. Better star catalogues were compiled (Halley attempted the first catalogue of the southern hemisphere from St Helena in 1677), and nebulae began to be studied.

Despite Newton's demonstration of a reflecting telescope, the difficulties in figuring a large mirror and finding a suitable material for it held back development until the middle of the eighteenth century, when a few up to about eighteen inches aperture were successfully made, of which the most famous are those of Sir William Herschel. Herschel also made a forty-eight-inch reflector of forty feet focal length, but found it very cumbersome to use, and generally preferred his smaller instruments.

Herschel conducted a systematic search of the heavens, found many more nebulae, and was able to resolve some of them into clusters of separate stars. In 1781 he discovered the planet Uranus. With larger reflectors of three-feet and six-feet aperture, the Earl of Rosse in the 1840s resolved many more nebulae, suggested that all were distant star clusters, and saw the ring and spiral nature of some. Again, however, these large telescopes were extremely unwieldy to operate, and weather conditions only occasionally justified the use of such a large aperture. One of Herschel's forty-eight-inch mirrors, and other historic telescopes, may be seen in the Science Museum.

The development of achromatic lenses in the middle of the eighteenth century, and better methods of producing large pieces of glass of uniform properties, resulted in larger refracting telescopes in the nineteenth century, culminating in the Lick thirty-six-inch refractor on Mount Hamilton in California (1888) and the Yerkes forty-inch refractor near Chicago (1897). Many of the famous photographs in the references were taken through these.

In the search for still larger apertures, the difficulties of making and supporting very large lenses (the forty-inch Yerkes lenses with their mount weigh half a ton) forced a return to reflectors—improved by this time by using a thin deposit of silver on glass as the reflecting surface. The sixty-inch Mount Wilson reflector was completed in 1908, the 100-inch in 1918, and the 200-inch on Mount Palomar in 1948. The details of the design, financing and construction of these large American telescopes are most interesting, and are given in King. They might form suitable subjects for individual reports, as might the latest news of the progress of the other large telescopes at present proposed or under construction.

Photography

These important techniques, which came into use in the middle of the nineteenth century, made possible the most exciting discoveries of more recent times. They were spectroscopy, the combination of telescope and prism to produce a spectrum of the light from different objects; photography, which permitted detailed and permanent records and long exposure observations; and photometry, which yielded really accurate comparisons of the brightness of different stars, and made temperature measurement possible. These are discussed in detail later.

Radioastronomy

The post-war development of radioastronomy depends on the transparency of the earth's atmosphere to two bands of radiation: radio waves as well as visible light. Other wavelengths, though emitted by the stars, are absorbed. A satellite-mounted telescope above the earth's atmosphere would have a much greater range of wavelengths from which to gain information about the emitter. Because of the far larger wavelength of radio waves, even the large modern aerial arrays have much poorer resolving power than light telescopes, and one of the problems of radio astronomy is to fix the position of radio sources with accuracy. There is an excellent film, *The Radio Sky,* available from A.E.I., which describes the development of radiotelescopes with many good views of them. Hoyle's *Astronomy* and Larousse are useful references.

Some students may not be familiar with the essentially similar nature of all parts of the electromagnetic spectrum. This can be discussed at this stage—most physics text-books give details. (The development of our understanding of electromagnetic waves is dealt with in topics II.7 and II.9.)

THE MEASUREMENT OF DISTANCES BY STELLAR PARALLAX

One of the recurrent problems in developing an understanding of the universe was the inability to detect stellar parallax. At first this was taken as evidence for a geocentric picture, but as the Copernican system became more widely accepted, and the diameter of the earth's orbit was more accurately measured, it appeared that the stars were such vast distances away that even using the 186,000,000 mile base-line obtainable by observing the position of a star at six month intervals, no difference in direction was detectable. There were clues, however, as to which stars might be the closest. Halley had shown, by careful comparison with earlier star positions, that the so-called fixed stars were not exactly fixed, but showed 'proper motions'—very small relative movements which over the centuries resulted in changes in the shapes of the constellations. On the assumption that the actual velocities of all stars were of the same order of magnitude, one which showed a large change in its apparent position was presumably closer than others.

Bessel selected the star 61 Cygni, which showed a large proper motion, and in 1838 succeeded in measuring a tiny parallax of about one-third second of arc, equivalent to a distance of about 600,000 astronomical units (1 astronomical unit = the average radius of the earth's orbit = 93,000,000 miles). Hence the distance was equal to about sixty million million miles, or eleven light years. Within a few months, Henderson succeeded in measuring the parallax of α-Centauri (the third brightest star in the southern hemisphere), and in 1840 Struve was successful with Vega. An idea of the smallness of the angles being measured may be obtained by equating them to the angle subtended by a small coin of a distance of about fifteen miles!

Despite the excitement of these first results, the extreme difficulty in detecting such small angles resulted in comparatively little further progress until the introduction of photographic techniques at the end of the century. Today about 5,000 of the closest stars have had their distances determined by this method, which reaches its limit at about 500 light years. For the ways of extending distance measurements beyond this range, see page 81.

THE APPLICATION OF GRAVITATIONAL THEORY

Gamow's *Gravity* is very helpful here. There are two striking examples of the application of gravitational theory during the nineteenth century. The first is the sequence of planetary discoveries. Herschel had found Uranus in 1781 during systematic observations. He at first thought it to be a comet, but a few months' observations showed it to be following a planetary orbit. Neptune was predicted in 1845 from discrepancies in the orbit of Uranus (orbital period eighty-four years), and identified soon afterwards by observation in the calculated position. Pluto, similarly predicted in 1915, when sufficient of Neptune's orbit (period 165 years) had been observed, was only discovered by systematic photographic searching in 1930, since the perturbations caused by it were too small to permit an accurate calculation of its position (Pluto's mass is less than 1/100 of Neptune's). The argument over who should claim priority for the discovery of Neptune throws an interesting light on the methods of the astronomers of the time, and provides an opportunity to introduce the modern problem of how to allocate observing time on the large telescopes.

Another example of a prediction based on gravitational theory and only later confirmed by observation is Bessel's prediction of the companion of Sirius. Sirius showed a comparatively large proper motion, but on careful observation this appeared to be along a wavy rather than a straight line, as Newton's laws of motion required. By 1844, Bessel had concluded that Sirius must be a double star, only one component of which was bright enough to be visible. The rotation of the two components about their common centre of mass, under the influence of their mutual gravitational attractions, gave to the visible component an apparently oscillatory motion. From the period and amplitude of this motion, and the measured distance of Sirius, which is one of the closest to the sun at nine light years, the masses of each component relative to the sun could be calculated. It was not until 1862 that the companion was observed, in the calculated position, by using a new twenty-inch refractor. Bessel made the same predictions about the bright star Procyon. Its faint companion was similarly identified in the calculated position, though not until 1892 when the thirty-six-inch Lick refractor had come into service.

Many other stars, perhaps one in ten in the universe, exist as binary or multiple systems. The application of gravitational theory, and the variation in brightness of those in which the components eclipse each other as they rotate, can provide very considerable information. Smart's *Some Famous Stars* is excellent on this.

SPECTROSCOPIC ANALYSIS

Early in the nineteenth century Fraunhofer, by combining a telescope and prism, analysed the light from different heavenly bodies. He mapped the hundreds of dark lines crossing the spectrum of sunlight, and showed that these occurred similarly in the light reflected from the moon and the planets, while some stars gave the same lines and others different ones. The explanation was provided in 1859 by Kirchoff, who identified particular groups of lines with known chemical elements, thus showing their presence in the sun, and requiring its outer layers to be at a high enough temperature to vaporise them, whilst the inner regions had to be hotter still. This is well described in Vaucouleurs (see also topic II.8). Spectroscopic analysis then developed very quickly. Stellar spectra showed the presence of chemical elements in the stars, small shifts in the wavelength of particular lines were explained as the result of stellar velocities (e.g. in a double star where the components revolve round each other), and the existence of new elements was postulated (e.g. helium in the sun, not identified on the earth until 1895). A method was now available for finding out more about the nature of stars than had been imagined at the beginning of the century. The Fraunhofer lines, and other emission and absorption spectra, can easily be demonstrated.

TEMPERATURE AND DENSITY MEASUREMENT

The surface temperature of a star affects its brightness, its colour, and the particular lines visible in its spectrum. Accurate photometric measurements of apparent brightness began about 1860, and had been extended to many stars by the turn of the century. Spectroscopic analysis succeeded in dividing stars into a number of classes depending on the lines visible in their spectra. By 1924, 225,000 stars had been classified in this way. In 1910, the first determinations of stellar temperature were made, and it became possible to relate surface temperature with a star's spectral class. Applying the general laws of radiation, better understood since the work of Planck and Einstein a few years earlier (see topic II.9), the apparent brightness of a star depended on its temperature, its surface area and its distance from the earth. Hence for those stars whose distance had been found, the measurement of apparent brightness and temperature, compared with the corresponding figures for the sun, permitted a comparison of the star's and the sun's surface areas. From these figures, the relative radii followed directly. Where the stellar mass could also be found, the average density of the star could then be calculated.

When applied to some stars, the method yielded surprising results. For example, the bright component of Sirius was found to be very like the sun—rather larger, hotter and more massive, but with a comparable density—but its feeble companion was found to have its mass, comparable with that of the sun, compressed into a sphere of only about 12,000 miles' radius, giving an almost unbelievably high average density about 65,000 times that of water. Subsequent measurements using other methods have confirmed this remarkable result (see p. 89). Other 'white dwarfs' have since been discovered. In such stars, the matter must be in a state very different from that of normal atoms as found on the earth. In contrast, the larger component of Epsilon Aurigae is found to have a radius of about 2,000 times that of the sun, a surface temperature of only about 1200°C, and an average density of only about a thousand-millionth that of water. Despite this immense range of size and density between different classes of stars, their masses are found to vary comparatively little—by a factor of only a few hundred times between the greatest and least.

TYPES OF STARS

Applying the methods just described, the characteristics of enough stars were determined to permit their rough division into 'dwarfs' and 'giants' by Hertzsprung in 1905. This classification was extended by Russell in the Hertzsprung–Russell diagram in 1913, where the main sequence dwarfs, red dwarfs, white dwarfs, red giants and super giants were distinguished. Larousse (ch. 14), deals with this, and has useful tables of the characteristics of a number of stars.

ENERGY MECHANISMS

What is the source of the energy radiated by a star? What happens as it is used up? What is the life cycle of a star as it condenses from a tenuous mass of gas? How does the sun compare with other stars, and what will be its probable future history? These and similar questions are discussed in popular fashion in Gamow's *Birth and Death of the Sun,* which also contains chapters explaining atomic structure and nuclear energy.

The main steps in the solution of problems concerning the constitution, energy production and evolution of stars may be summarised as follows:

(a) Helmholtz's contraction hypothesis (1854) assumed that the sun had been contracting from a tenuous cloud of cool gas, under the action of the gravitational forces of attraction between molecules. This contraction causes a rise in temperature, and the energy radiated comes from the original gravitational energy of the separated molecules. To produce the present rate of radiation, the sun would only have to contract by about two kilometres per century—but the total energy which would have been liberated in the contraction from even an infinite size down to the present dimensions can be calculated, and is found to be far less than that required by other considerations. Whilst gravitational contraction is certainly part of the source of stellar energy, it is by no means the whole.

(b) The discovery of radioactivity by Becquerel in 1896 (see topic II.8, p. 170) provided another possible mechanism for energy liberation—if radioactive elements were present in the stars. The radioactive elements first discovered on the earth were heavy elements, however, and studies during the 1920s showed that the stars were predominantly composed of the lightest elements, hydrogen and helium.

(c) The Hertzsprung–Russell diagram (1913) suggested ways in which the luminosity and temperature of a star might vary during its evolution.

(d) Between 1916 and 1924, Eddington developed the theory of the stellar interior, according to which stellar matter is in equilibrium between the expansive forces of gas and radiation pressure, and the gravitational force of attraction. From the theory, Eddington was able to estimate the temperature in the interior of stars, and obtained very high values—of the order of tens of millions of degrees.

(e) The discovery of the artificial transmutation of elements by Rutherford in 1919 led to much work on the radioactivity of isotopes of the light elements which might exist in the stellar interiors.

(f) From 1929 onwards the idea became accepted that thermonuclear reactions, in which some part of the atom's nuclear mass was converted into energy, produced the vast quantities of energy radiation by a star during its lifetime. The cycle of reactions which best fitted the observed rate of production was worked out in 1938 by Bethe and Weizsäcker. It was then possible to suggest a life cycle for a star based on: *condensation* from tenuous gas, mostly hydrogen, with a small proportion of heavier elements; and *gravitational contraction,* producing a temperature increase great enough in the interior for the *thermonuclear reactions* involving the conversion of hydrogen into helium to take place, and leading to an *equilibrium size*. As more of the hydrogen is converted into helium the temperature rises, and the rate of energy production increases (the sun should be at its hottest in about ten thousand million years' time). Thereafter, as the supply of hydrogen is exhausted, the radiation pressure from within falls, the star contracts and cools, ending as a dwarf star of high density. The process is vividly described in Gamow and in Hoyle's broadcast talks on *The Nature of the Universe*.

FURTHER DISTANCE MEASUREMENTS

The measurement of stellar distances was greatly extended by a spectroscopic method introduced by Adams in 1916. He observed that the intensities of different lines in a star's spectrum varied with its brightness. The absolute brightness of the closest stars had already been calculated from their apparent brightnesses and their trigonometrically measured distances. A curve relating absolute brightness and the ratio of the intensities of convenient spectral lines could thus be plotted. From this, the absolute brightness of more distant stars could be inferred once their spectra were obtained; and from their apparent brightness their distances could be calculated. This method, known as spectroscopic parallax, is subject to inaccuracies and, like the other methods to be described, relies upon the assumption that physical behaviour at greater distances from us is the same as that close at hand—but with these reservations it has permitted the extension of distance measurement from a few hundred to several thousand light years, and the number of stars with known distances has increased to many thousands.

One particular class of variable stars has provided a method of extending distance measurement still further. The Cepheid variables, called after their prototype Delta Cephei, are pulsating stars which vary in brightness with periods of days. It was first discovered in 1912 that there was a relationship between the period and the average apparent brightness of these stars, but the distances of even the nearest of them are large, and can only be measured with uncertainty. The conversion of apparent to absolute brightness is therefore also uncertain, and distances based on earlier figures have had to be revised. In principle, however, the absolute brightness can be obtained from the star's period, and its distance inferred from its apparent brightness. Cepheid variables have been identified in star clusters, and, subject to the uncertainty mentioned, have given their distances up to about 100,000 light years. From these measurements, the dimensions of our galaxy have been deduced (see, for example, Smart, ch. VI: Larousse, ch. 13).

THE DISTANT NEBULAE

The first catalogue of nebulae was published in 1784 by Messier, who listed 103 objects. His classification is still used (e.g. M31 = great Andromeda nebula—the only one visible to the naked eye in the northern hemisphere). The Herschels added much to this, and the *New General Catalogue (NGC)* was published by Dreyer in 1890. The real break-through came in the 1920s with Hubble's use of the Mount Wilson 100-inch reflector. This resolved many nebulae into separate stars, amongst some of which Cepheid variables were identified. This made possible the measurement of their distances, and hence the division into the very distant extra-galactic nebulae millions of light years from us, and the closer galactic nebulae, mostly star clusters and luminous gas clouds. Hubble also divided nebulae into elliptical and spiral categories (a few fit neither). There are many excellent photographs in Larousse and Hoyle's *Astronomy*.

The Cepheid method cannot be applied to the more distant nebulae, whose distances have to be estimated by assuming either that the brightest stars in each have the same absolute brightness, or that the nebulae as a whole have. From their distances, nebular dimensions can be determined. The extra-galactic nebulae are found to be comparable in size to our own galaxy.

In 1929 Hubble announced the remarkable finding that the more distant a nebula was, the faster it appeared to be receding from us. The velocities of recession were measured by the apparent shift of known spectral lines towards the red end of the spectrum—the Doppler effect, discussed in Advanced level physics textbooks. The values ranged up to about one-seventh of the velocity of light at hundreds of millions of light years' distance. Since then, the use of the 200-inch telescope has extended the range of observations up to about 2,000 million light years, and some extremely distant objects have been identified with red-shifts corresponding to velocities of nearly half that of light. Whether this Doppler explanation is correct remains a matter of conjecture, though no alternatives have gained acceptance. The apparent expansion of the universe has been discussed by many writers. It is one of the principal observations to be explained in any theory of its creation.

THE AGE AND ORIGIN OF THE SOLAR SYSTEM

This topic is well covered in Smart's *Origin of the Earth,* which students could study for themselves and continue from recent references, e.g. Larousse, Hoyle's *Astronomy* and the scientific periodicals. Summarising briefly:

The age of the earth

(See *The Discovery of Time*)

Seventeenth century: Archbishop Usher—4004 B.C.

Nineteenth century: Lord Kelvin—the shape of the earth's crust on solidification, the rate of solar radiation and the rate of cooling of the earth all suggested about 100 million years. This was in conflict with the geological evidence: the rate of accumulation of salts in the oceans and the rate of laying down sedimentary rocks suggested a far greater age.

Twentieth century: (a) The study of changes in the earth's crust, work on radioactivity as a source of the earth's heat and the discovery of possible thermonuclear reactions in the sun, have lengthened Lord Kelvin's estimates, and brought them more into line with geological and other values. (b) From the study of the rate of radioactive decay of uranium and thorium to lead, values of about 4,000 million years for the earth's crust are obtained. (c) The moon is slowly receding from the earth as a result of tidal friction. The two can be calculated to have been in contact about 4,000 million years ago.

The origin of the solar system

There have been numerous theories since Laplace's nebular hypothesis (1796). These are well summarised and discussed in Smart's *Origin of the Earth,* though this needs supplementing with more modern information such as Hoyle's *Astronomy* or *Frontiers of Astronomy*.

The principal observation, that all the planets (and most of their satellites) rotate in the same direction round the sun and in approximately the same plane, suggests a common origin for them all. Successive theories have flourished as knowledge of the nature of the sun and the planets has increased. When the presence of the majority of chemical elements had been identified in the sun by spectroscopic means, it seemed likely that the sun and planets had condensed from the same material (Laplace's hypothesis), or that the planets had come from the sun (Jeans' tidal theory, etc.), though the mechanics of the processes presented considerable difficulties. When further work showed that the sun contained mainly hydrogen and helium, with only a small proportion of the heavier elements, these ideas looked less likely, and double star theories or condensation from separate dust clouds (Weizsäcker) received support. Recent work on the composition of the outer planets, and on the possible magnetic forces within a rotating star, have shown how the rapidly rotating sun might have flung out the material to produce our planetary system. The subject is not yet closed, but on the type of mechanism postulated depends the interesting question of the number of other planetary systems similar to our own in the universe. On the assumption of solar rotation throwing off planetary material, the answer could be very large indeed—Hoyle puts it in the thousands of millions, since this process could be part of the normal behaviour of many stars. Weizsäcker's mechanism should also be of common occurrence. On the other hand, the chance close approach of two stars is very unlikely.

On modern theories, other planetary systems would be very like our own, so the conditions for life may exist in many other parts of the universe, though separated from communication with us by millions of light years. At this stage it is interesting to review the conditions necessary for life: not too extreme a temperature range, suitable elements, suitable atmosphere. The importance of the escape velocity of different molecules in determining the latter is discussed in *The Origin of the Earth* (chs. III and IV). A very useful article appeared in *The Scientific American* (April 1960). Recent photographs of the earth from rockets and satellites illustrates how difficult it is to deduce the existence of *intelligent* life on a planet from the patterns on its surface, even from comparatively low altitudes. At present, attempts are being made to discover whether any of the radio waves reaching us from distant sources show signs of an intelligent origin.

THE ORIGIN OF THE UNIVERSE

The October 1966 issue of *Science Journal,* entitled 'The New Universe', contains a very useful summary of modern work. Hoyle's *Galaxies, Nuclei and Quasars* (1965), though complicated in places, gives a very valuable insight into recent thinking. The 'big bang' hypothesis follows from the recession of the nebulae (if correctly interpreted), and suggests an age of about 10,000 million years, when all matter was concentrated in the 'primaeval atom'. The steady state theory is well described in the last chapters of *Frontiers of Astronomy,* and later developments, including the idea of an oscillating universe, are discussed in *Galaxies, Nuclei and Quasars.* This is a field in which ideas are extremely fluid, and one can see different theories changing rapidly. (Compare for instance Hoyle's *Nature of the Universe* (1950) with his later books.) The more modern references bring out the vital role of radio astronomy in observing very distant objects, and hence in providing much of the information necessary to discriminate between these different theories.

Students may question whether the expense of further research in this subject is justified. This may form the subject for a useful final discussion—if in fact it has not been raised earlier in the course!

LUNAR AND SOLAR CALENDARS

The moon's period is on the average 29·53059 days; a solar year is 365·24220 days. Two difficulties therefore have to be overcome in constructing a lunar calendar. Firstly, the moon's period is not an exact number of days, and observations of the moon will give alternate months of twenty-nine and thirty days (with occasionally two consecutive thirty-day months due to the slight excess of the period over twenty-nine and a half days). Secondly, the normal lunar year of six 'hollow' and six 'full' months contains only 354 days and falls behind the sun at the rate of eleven days per year. If the two calendars are to be kept in step, an extra month has to be inserted every third year, and sometimes oftener. The calendar problem involves finding the regular sequence that will achieve this.

From our modern knowledge of the exact periods, we can calculate the larger period which is a common multiple of the month and the year: after this time the sun and moon will return to the same mutual positions. An exact common multiple does not exist, but various approximations to the ratio 29·53059:365·24220 may be found. The simplest is 8:99, indicating that eight solar years are nearly equal to ninety-nine lunar months (actually 2921·94 and 2923·53 days). Eight normal years would contain only ninety-six months, so three years must have the thirteenth month included (say the third, fifth and eighth). This system, the octaëteris, was discovered in early Greek times, but it is not a very good approximation; after only twenty-four years the moon date will be five days late relative to the sun. Far more precise is a period of nineteen years which contains 235 months (6939·60 and 6939·69 days); a thirteenth month must here be intercalated seven times (say in the third, fifth, eighth, eleventh, thirteenth, sixteenth and nineteenth years) and the cycle must contain 110 'hollow' months of twenty-nine days and 125 'full' months of thirty days. To spread the 110 days missing from a uniform calendar of 235 thirty-day months as evenly as possible, every sixty-fourth day is omitted—thus determining which become the twenty-nine-day months.

The nineteen-year cycle was discovered by the Athenian astronomer Meton about 433 B.C. It appears in late Babylonian records about 380 B.C. Meton's nineteen years containing 6940 days give an average length for the year of 365 5/19 days: about thirty minutes too long. A correction was applied by Callippus in 330 B.C., who took a cycle of $4\times19=76$ years containing 27,759 days ($4\times6940-1$). This gives a year of exactly $365\frac{1}{4}$ days, and a lunar month only twenty-two seconds too long. (Note that $5/19 = 1/4 + 1/76$, an interesting example of the usefulness of the unit fraction method.)

A further correction by Hipparchus about 125 B.C. took a cycle of $4 \times 76 = 304$ years containing 111,035 days $(27{,}759 \times 4 - 1)$. This gives a year which is only about six-and-a-half minutes from the true mean tropic year, and a lunar month less than one second from the accepted modern value! (See Heath, ch. XIX.)

NAMING OF STARS

The brightest stars have separate names, either Greek (e.g. Castor and Pollux) or Arabic (e.g. Algol, Altair, Betelgeuse, Deneb, Rigel, Sirius, Vega). A systematic nomenclature was introduced by Bayer at the beginning of the seventeenth century; the brightest star of a constellation was given the first letter of the Greek alphabet, the second brightest β and so on. Hence α-Orionis = Betelgeuse, β-Orionis = Rigel, etc. After exhausting the Greek alphabet, the Roman one was used: a-Orionis, b-Orionis, etc. When the telescope revealed many more stars in each constellation, Flamsteed introduced a further numerical system: 1-Orionis, 2-Orionis, etc. Today the fainter stars are identified from their numbers in a particular catalogue, or by their coordinates on a photographic plate.

PRECESSION OF THE EQUINOXES

The direction of the earth's axis of rotation does not remain constant with respect to the fixed stars, but has a conical motion similar to that of a spinning top. At present the axis is pointing towards the brightest star in the constellation of Ursa Major—hence now known as the Pole Star. But its direction is slowly tracing out a circular path on the background of the fixed stars, with a recurrent period of 26,000 years. This fact was first noticed by Hipparchus, when he compiled his famous catalogue of the positions of over 1,000 stars, and found a noticeable shift in their apparent positions (about 2°) compared with observations recorded 150 years previously. From careful measurements, Hipparchus was able to identify the cause of this shift. He realised that it would slightly affect the time for the sun to appear to return to the position giving an equinox—hence the name 'precession of the equinoxes'—and that it was therefore important to distinguish the length of a year, based on observations of the time from one equinox to the next (the tropical year), from that based on the apparent motion of the sun against the background of the fixed stars (the sidereal year). This is discussed in Berry, the section on Hipparchus.

The explanation of this 'wobble' of the earth's axis of rotation was only given some eighteen centuries later by Newton. Due to its rotation, the earth's shape is not perfectly spherical, but bulges at the equator. The gravitational attraction exerted by the sun on this 'bulge' tries to turn the earth's axis perpendicular to the plane of its orbit. But, like any spinning body, the response to this turning effect is not a simple adjustment of the direction of the axis, but the precession seen for instance when an attempt is made to change the axis of a spinning top. (See Gamow, *Gravity*, ch. v.)

STELLAR MAGNITUDES

Estimates of brightness were first made by Hipparchus, who designated about twenty of the brightest stars as 'of the first magnitude', the next fifty or so as 'of the second magnitude', and so on until the sixth magnitude stars were just visible on a clear night.

Stellar magnitude is now defined in such a way that a star which appears to be two-and-a-half times brighter than another has a magnitude of one less. On this basis, the relative brightnesses of the first six magnitudes will be 100, 40, 16, 6·3, 2·5, 1. A few very bright stars have magnitudes less than unity, e.g. Vega with 0·14 and Sirius, the brightest star in the sky, $-1·52$.

This scale covers an enormous range of brightness. With the 200-inch telescope, stars as feeble as the twentieth magnitude can be seen, and weaker ones can be photographed with long exposure times. The range is therefore about twenty-four magnitudes, corresponding to a brightness ratio of 4,000,000,000 to 1.

RELATIVITY

At some stages in this course, questions about the theory of relativity come up. It is not easy to give students without the necessary mathematical background a real understanding of this. There have been numerous attempts to present the theory in non-mathematical terms, but these can give a confused and superficial impression. Bondi's recent *Relativity and Common Sense*, however, introduces the subject in a way that students of a course such as this might understand. *Relativity for the Layman* is another possible reference, and the B.B.C. published a very useful pamphlet, 'Relativity Today', in 1963.

The principal places where the consequences of the theory arise are:

the interconvertibility of mass and energy—as an explanation of the source of energy in the thermonuclear reactions in stars;

the variation of mass with velocity—as applied to the high velocities of the receding nebulae;

the problem of the orbit of Mercury;

the effect of a strong gravitational field on light, as shown by the deflection of light from a star as it passes close to the sun (Eddington, 1919), and by a change in the frequency of spectral lines emitted from very dense bodies (Adams on Sirius B, 1925).

GLOSSARY

A number of astronomical terms used in the text are explained here:

The *celestial sphere* is an imaginary sphere of which the earth is the centre, and on the surface of which the heavenly bodies appear to be situated. The point directly overhead for an observer is the *zenith*. The position of a heavenly body may be specified by its *altitude* (its angular distance above the horizon) and its *azimuth* (measured along the horizon in degrees west of south). A *great circle* is a circle whose centre is at the centre of the sphere. The north-south great circle passing through an observer's zenith is his *meridian*.

The *ecliptic* is the great circle on the celestial sphere which the sun appears to trace out in a year. It is therefore also the circle in which the plane of the earth's orbit cuts the celestial sphere. The angle between the ecliptic and the equator is called the *obliquity of the ecliptic*. The points at which the ecliptic cuts the equator are the *spring and autumn equinoxes*.

The *zodiac* is the zone on either side of the ecliptic in which the planets appear to move. The zodiac is divided into twelve *signs* or constellations, each occupying 30° of the ecliptic. These are: Aries (ram), Taurus (bull), Gemini (heavenly twins), Cancer (crab), Leo (lion), Virgo (virgin), Libra (scales), Scorpio (scorpion), Sagittarius (archer), Capricornus (he-goat), Aquarius (water-carrier), Pisces (fishes).

At *conjunction*, the earth, sun and planet are in line. *Superior conjunction* occurs when the planet is on the far side of the sun; *inferior conjunction* (Mercury and Venus only) when the planet is between the sun and the earth. *Opposition* (not Mercury and Venus) occurs when a planet is in line on the far side of the earth from the sun. Because of the relative motions of the earth and the planets, a planet appears to advance through the signs of the zodiac, then turn back and subsequently recommence its forward movement. The turning back is known as *retrograde motion*.

An *occultation* occurs when the moon passes in front of another heavenly body.

Culmination is when a star passes across the meridian.

Heliacal rising is when a star rises above the horizon at the same time as the sun.

II Matter theory

DAVID STEELE

The topics in this section follow quite closely the development in *The Architecture of Matter* by Toulmin and Goodfield; and for an introduction to the problems of matter theory one cannot improve on their introduction (pp. 17–23).

Notice particularly the contrast between astronomy and matter theory and the difficulties experienced in formulating problems in the latter. This confusion, which has persisted throughout the development of matter theory, makes difficult the decision as to what are the significant stages in the story; it is a great temptation to include only material which forms a logical development up to the present state of knowledge. Other 'less relevant' stages, such as alchemy and the seventeenth century, should find a place in a balanced course on ideas. Another danger in the retrospective approach to matter theory is that of anachronisms: several familiar terms, particularly *atom* and *element,* have a long and complicated history and it is important that their meaning at any stage is clearly understood and not distorted by modern connotations.

II.1 Ancient civilisations and materials

Archaeological remains of the time before 5000 B.C. enable us to draw only a fragmentary picture of the technology of the period; such information as we have seems to indicate a very limited use of materials 'ready to hand', such as flints, wood, clay (which was baked to make crude pots), pigments such as malachite, and (later) gold and copper.

After 5000 B.C. certain of the nomadic tribes settled in the fertile areas around the great river valleys in China, India and around the Mediterranean Sea. We are concerned here with Western civilisations which began to develop in the valleys of the Nile, Euphrates and Tigris; areas around these valleys constitute the 'fertile crescent'. This is shown in map I.1 (p. 16). For a chronological outline of early civilisations see table I.2 (p. 19).

The political structure of these early communities was complex and varied but one fairly common feature was the secrecy which surrounded the spread of knowledge at all levels in society. Workers in metals, glass, pigments, etc. (the craft tradition), jealousy guarded the secrets of their crafts and passed them on only to the initiated. One reason for this secrecy was, of course, to protect their interests against rivals; but as the secrecy deepened and the initiation ceremonies became more eleborate, the discoveries themselves acquired mystical and religious significance and this had important effects on alchemy. A similar fate befell the 'higher' knowledge, particularly mathematics and astronomy. Such knowledge had, since earliest times, been in the hands of the priests. (For more details of the development of early civilisations see Glover and Childe, chs. 7–11.)

METALS

Before 5000 B.C. metals were not much used; gold and copper are the only ones found in the remains of this age and gold was probably more common than copper. After 5000 B.C. the variety of materials and skill in their use gradually increased. This is summarised in table II.1, which gives details of metals and other materials requiring skill in applied chemistry for their preparation. Materials found only very rarely are enclosed in brackets.

Table II.1
The discovery of materials in the ancient civilisations

B.C.	*Egypt*	*Mesopotamia*	*Aegean*, etc
4000	Predynastic: gold, silver, lead, copper, glaze (glass).	Sumer-Akkad: *Lagash, Kish, Ur*: gold, silver, lead, (iron), copper, tin–bronze, glaze, (glass).	
3400	I Dynasty:		
3000	(tin–bronze).	lead and antimony bronzes, metallic antimony.	Early Minoan: *Knossos*: copper, gold, silver, lead, tin–bronze, glaze, purple dye.
2000	XII Dynasty: (tin), iron tools.	tin–bronze common.	*Mycenae*: glass.
1700–1500	XVIII Dynasty: tin–bronze common, tin, glass factories, cobalt, glass, indigo, useful iron.	useful iron, steel weapons, tin oxide glaze, cobalt glass.	The Hittites: iron working near Black Sea. *Tiryns*: lead–copper glass.
1350–1200			*Palestine*: zinc brass
1000	Nubian iron industry, Assyrian iron common in Egypt.	Assyrian iron common.	The Phoenicians: traffic in tin. Iron Age: Classical Greece.
650		Assyrian chemical tablets, imitation gems.	

From about 4500 B.C. copper became the most common metal and we must assume that furnaces had been devised which would enable copper to be obtained by the reduction of its common ores. This was probably the first, and certainly a most important technical advance in metallurgy; man no longer relied on the metals to hand in nature but was able to produce them from more abundant materials. A little later bronze objects appear. It is thought that early bronze was made, not by melting together copper and tin, but by smelting the ores together; in some ores the metals occur together. The greater mechanical strength of bronze was soon appreciated and, although the first bronze may have been the result of the chance mixing of the ores, it is clear from analyses of objects of the period that the ancient metallurgists controlled the proportion of the two metals to give the alloy with the best mechanical properties (about nine per cent tin). (See Stillman ch. 3 where several analyses are given.) This constancy in the composition of bronzes from a wide area and from over a long period of time seems to indicate that men must have experimented with mixtures and found by trial and error the composition that best suited their purpose, and having found such a composition, endeavoured to keep it in order to have uniformly reliable products.

As the design and efficiency of furnaces improved, so other metals such as lead and mercury appeared. Finally, in about 2000 B.C., furnace techniques were sufficiently advanced for iron to be produced fairly commonly.

OTHER MATERIALS

Although metals represent the most important technical advances in this period, a similar improvement can be observed in glass, pottery, pigments, medicines, etc. In these cases as in metallurgy we find not only new and improved techniques but also control over the composition of materials. For example, the beautiful colours produced in glass by traces of metals (cobalt gives an intense blue) were known to the Egyptian glass makers, and the consistent quality of their colouring shows they controlled most carefully the tiny quantities of colouring materials used.

THE CRAFT TRADITION

In these ancient civilisations we see the beginnings of the craft tradition which has developed alongside matter theory, often furnishing valuable experimental data for the theorist and yet itself receiving very little useful help in return. Possibly only during the period of alchemy and in this century has theory really helped the development of the practical arts (see *Architecture*, ch. 1).

EXPERIMENTS AND DEMONSTRATIONS

(a) If available, samples of minerals containing native metals such as copper and gold can be shown. These can be compared with minerals containing metals in the combined state, such as malachite and lapis lazuli; from the bright colour of these latter minerals their early use as pigments and for ornament can be appreciated.

(b) Strips of copper and of bronze of similar dimensions will show the greater toughness of the latter.

(c) It is possible to demonstrate the reduction of some ores to metals—see elementary chemistry textbooks and *Nuffield Basic Chemistry Course* (pp. 191–4). The 'charcoal block' reductions in qualitative analysis are examples of the process.

(d) The local museum may have some interesting material by which to illustrate this topic; the Science Museum in Kensingon certainly has.

(e) The problem of the accurate dating of the remains of the ancient civilisations provides an opportunity to discuss the application of scientific techniques to archaeology—not only dating but also the preservation of antiquities (see Brothwell and Higgs, *Science in Archaeology*).

II. 2 Greek matter theory

The Greek philosophers believed in 'eternal principles' capable of *rational* expression; that is, in the essential reasonableness of nature. Until this approach was adopted man's knowledge, though extensive, remained disjointed and purely empirical. After the Greeks there was a logical framework on which discoveries and observations could be based, and thereby new and often profitable patterns were detected (see Sambursky, ch. 1).

It is no discredit to the Greeks that much of their framework, in science anyway, has had to be renewed. They were the first to show the need for such a framework and showed how best it is to be constructed. Herein lay their great achievement. This will help counter the objection often made that courses on Greek science are irrelevant today. Another most important answer to this objection is that in the science of every age it is the questions asked rather than the answers offered which are important. The Greeks posed a series of fundamentally important questions and the subsequent history of science (and philosophy and politics) is essentially the history of man's attempt to answer them and to make sense and use of his answers.

THE EARLY PHILOSOPHERS

Though our knowledge of the earliest Greek thinkers is fragmentary, it is generally agreed that Greek rationalism originated in Ionia and in particular in Miletus. Miletus was an important trading port and constant political and intellectual friction in this cosmopolitan community made it a breeding ground for new ideas. Not only was there much fruitful exchange of ideas and information, but institutions and habits were compared. Comparison led to judgement and judgement to reflection; and reflection on similarities and differences probably deepened to reflection on permanence and change, the first and most fundamental problem of science in general and matter theory in particular.

Conservation

Among the earliest thinkers to show this rational approach to problems was Thales (624–546 B.C.) of Miletus (see map I.1, p. 16). It seems that he was the first to state clearly and face the most important problem of *conservation*. Though the world is constantly changing, we are instinctively aware of a continuity, of something being conserved. What is it that is conserved? Those who have studied physical science will already be familiar with the conservation of mass and of energy and will appreciate that contemporary science is still very much concerned with the problem.

Aristotle summed up the thinking of the early Ionian philosophers in his *Metaphysics* (Ross ed., O.U.P., p. 933):

Of the first philosophers, then, most thought the principles which were of the nature of matter were the only principles of all things. That of which all things are consist, the first from which they come to be, the last into which they are resolved (the substance remaining, but changing in its modifications), this, they say, is the element and this the principle of things, and therefore they think nothing is either generated or destroyed, since this sort of entity is always conserved. For there must be some entity—either one or more than one—from which all other things come to be, it being conserved. Yet they do not all agree as to the number and the nature of these principles. Thales, the founder of this type of philosophy, says the principle is water.

The evidence by which Thales supported his answer is characteristic of the Greeks: reasonable interpretation of common observations. Here are two examples. Water is the only common substance we meet as solid, liquid and gas, therefore it is not unreasonable to suggest that it might lie behind this most obvious classification of materials. Secondly, Thales is supposed to have been very impressed by the effects produced by the annual flooding of the Nile (he had travelled in Egypt). Desert wastes are —albeit temporarily—turned into fertile land; and water itself seems to produce an abundance of living things. Thus the answer 'water', though it may seem hopelessly inadequate to us, did account quite satisfactorily *at that time* for the chief features of man's environment. Although the new way of thought which we see illustrated here was ultimately to displace the mythological explanations characteristic of earlier writers, it is unreasonable to suppose that the powerful grip of such explanations was released overnight. Thales was probably still influenced by mythology and from this might have derived his answer water, for 'Okeanos' in

Homer was the source of Gods and all things. What is important is that he attempted to substantiate his answer by evidence which, however inadequate and far-fetched it may *now* seem, was recognisably scientific.

Anaximander (611–545 B.C.), a disciple of Thales, objected that as an answer to the conservation problem, water was too simple and too obvious. A substance fulfilling such a fundamental function must be capable of infinite possibilities; it must have no specific properties or attributes. As a name for the primary matter of the universe, Anaximander suggested 'apeiron', which means boundless, indefinite or infinite. The properties of actual materials comes from a separation of opposites, such as hot and cold, dry and wet, contained in apeiron. These qualities are constantly increasing and decreasing and this accounts for change (see *Architecture,* p. 50). Thus Anaximander introduced an important refinement to the problem of conservation by recognising that it was essentially the problem of permanence and change (*Architecture,* pp. 46–8). Apeiron is permanent but the qualities in it vary and thereby give rise to change. This clarification of the problem of conservation later stimulated most fruitful answers.

A disciple of Anaximander, Anaximenes (*d.* 525 B.C.), attempted a compromise between the ideas of Thales and Anaximander. He felt that this fundamental stuff must have a more meaningful name than apeiron yet must be less specific than water. He suggested air; it is invisible yet demonstrable (in the wind), infinite yet collectable (this was clearly shown by Empedocles, see p. 101). Air is permanent and changes occur by rarefaction and condensation. Anaximenes also expressed the idea that air is the breath of the universe like the breath in the body. This was later combined with 'fire' (from Heraclitus) and developed into the powerful explanatory concept 'pneuma' by the Stoics (see p. 109). Another important feature of this theory of Anaximenes was that rarefaction and condensation of the air presuppose motion and this was the first attempt to suggest a *mechanism* of change. Now, motion is describable in mathematical terms, thus this represents the first move towards the expression of qualities in terms of quantities, a technique which later led to rapid and fruitful advance and which became a characteristic of the 'scientific 'approach.

It is worth noting that these three thinkers sought a *single* fundamental substance. Later theories involved more than one substance, but it has always been an instinctive desire of the scientist to explain the maximum of phenomena by the minimum of hypotheses—ideally one. (Think of the initial satisfaction felt at the beginning of this century when it was thought that the complexity of ninety or so elements had been reduced to three 'fundamental' particles—electrons, protons and neutrons, and of the present unrest that the number of 'fundamental' particles now numbers over two hundred.) The principle of economy of explanation has always been highly prized.

Permanence or change?

The important problem of permanence and change was further intensified and clarified by Heraclitus of Ephesus (540–475 B.C.) and Parmenides of Elea (*c*. 500 B.C.).

Heraclitus considered change to be the essence of reality; things which did not appear to be changing were in a state of dynamic equilibrium. The most spectacular changes were wrought by fire; also the similarity between fire and life was clearly appreciated by several Greek thinkers. Fire occupied a central position in the thinking of Heraclitus, though he did not regard it quite as Thales regarded water.

It is an ever-present danger to read into Greek science relatively sophisticated modern concepts, and it would be foolish to credit Heraclitus with the discovery of energy or thermodynamics; nevertheless it is clear that he did appreciate in a crude and speculative way some of the fundamental principles involved. For example, in emphasising change he considered that a world in which no changes occurred was inconceivable and that the universe might end thus when all possible changes have occurred. This is somewhat akin to the 'heat death' predicted from thermodynamics (see Ubbelohde, *Man and Energy*, ch. 13, and *Time and Thermodynamics*). For more on Heraclitus, see Sambursky (pp. 198–9) and Kirk and Raven (pp. 196–200).

Since awareness of change must be through the senses, the teachings of Heraclitus stressed the importance of sensory perception. Reality is that which is perceived; little is to be gained by seeking a deeper reality beyond the senses.

Parmenides and his more famous pupil Zeno taught precisely the opposite. For them change was illusory, the senses were fallible so the world thus revealed was an illusion. Reality was in the mind and only the products of reason were permanent and reliable (see Sambursky, especially pp. 147–57, for an excellent discussion of this, including Zeno's famous paradoxes). In the writings of Heraclitus and of Parmenides we see posed clearly the problem of the reliability of sensory perception; closely related to it is the relationship between mind and matter. These are further examples of the fundamental problems first posed by the Greeks, problems which have exercised the minds of men ever since.

The problem of permanence and change was faced and answered in two quite different ways by Empedocles (500–430 B.C.) and by Democritus (460–370 B.C.). Their two answers, which for nearly two thousand years were irreconcilable, formed the twin pillars on which chemistry was built. Only at the beginning of the nineteenth century were they wedded into a single theory by Dalton (see p. 138).

The element theory

The theory of Empedocles is a development of the earlier ideas of Thales and his successors. Empedocles suggested that the permanence in the world is due to four fundamental materials—the four elements (Empedocles was the first to use the word in this way). The four elements were water, air and fire (suggested by his predecessors) to which Empedocles added earth. The changes observed in nature are caused by the mingling and separation of the elements which themselves undergo neither change nor disintegration. The agencies responsible for the mingling and separation of the elements were love and hate. (It is arguable that the modern counterparts of these, attraction and repulsion, are only superior by virtue of the precision with which they can be expressed.) It is interesting to note that the four elements consist of materials representing the three states of matter together with the agency by which they are interconverted. This perhaps shows an intuitive sense for what is really significant in the material world.

Empedocles' theory of material as an organised mixture of different stuffs, themselves irresolvable into anything simpler, formed not only a rational explanation of matter but directed subsequent enquiry along a fruitful path. It found an important place in Aristotle's scheme of things and thereby remained influential for two thousand years.

The first experiment

Empedocles is important in that he was the first to give an account of a recognisably 'scientific' experiment. In his attempt to explain the mechanism of breathing, Empedocles demonstrated the corporeal nature of air—hitherto, and often subsequently, identified with the void—by a simple experiment with a water-clock. He showed that if the air-hole is blocked then water will not enter when the vessel is immersed in water; when the finger is removed from the hole, then water rushes in and displaces the air (see experiments, p. 115).

The importance of this experiment is two-fold: first, that it is the earliest recorded experiment—the demonstration of an hypothesis ('air is a body') by means of a specifically stated sequence of operations ('finger on, water kept out; finger released, water enters, etc.'). Secondly, and perhaps more important, this experiment shows that matter can exist in a form too fine to be apprehended by the senses, yet which can exert considerable power (holding back the water). This enables us to discover truths which are not directly observable, by inference based on observation. This technique was of the utmost value to the atomists.

Atomism

Atomism was the other attempt to explain permanence and change and endeavoured at the same time to incorporate number into nature—an aspect which appealed greatly to Plato. The theory is supposed to have been first propounded by Leucippus (500 B.C.) and developed by Democritus (460–370 B.C.). There are three sources of information on Greek atomism which enable us to get a clear and fairly complete picture of the theory. The first is Aristotle who, unable to accept the atomism of Democritus or to find a place for it in the system of the universe, gave a very full account of it in order to refute it adequately. The second is Epicurus' *Letters* and the third, and most readily available source, is *De Rerum Natura,* a didactic poem in honour of the Epicurean philosophy (which was based on atomism) by the Roman poet Lucretius (*d.* 55 B.C.).

It is probable that Leucippus propounded physical atomism as a counter to the paradox of division posed by Zeno in his paradoxes. Since there is no limit to mathematical division then the number of points between two given points is infinite. Now if this is applied to physical division then matter is divisible into smaller and smaller bits and in the limit into nothing and thus it can hardly be said to be conserved. Leucippus therefore suggested that there is a limit to the physical division of matter; this limit—the smallest particle into which matter can be split—he called an *atom*.

This idea was developed into the atomic theory by Democritus. The chief postulates of his theory were:

(a) matter is conserved, it cannot be infinitely divided, the smallest particles are atoms which can neither be created nor destroyed (Epicurus added that they were invisible and introduced 'weight' as a cause of their motion);

(b) there are as many kinds of atoms as there are materials;

(c) atoms of different materials differ only in size, shape and movement;

(d) atoms move in a void.

The supporting evidence for this theory is typical of Greek science—reasonable interpretation of everyday observations: the wearing of a step by pilgrims, where we do not see each increment of wear yet the hollow in the step is the result of many small increments; the fact that an array of individual moving objects, such as a flock of sheep or a column of soldiers, if viewed from a sufficiently great distance seem to lose their individuality and appear as a uniform area to the eye (justification for our not being able to see individual atoms in matter); the invisible wind uprooting trees. All these are examples of the reality of invisible things being inferred from their visible effects.

Perhaps the most striking example of the ability of these Greek thinkers to give penetrating if speculative explanations of common phenomena is Democritus' account of the movement of specks of dust observed in a sunbeam. He gave precisely the same explanation that we now give for the Brownian motion—that the movement of the specks is due to the continual and random bombardment by the atoms of air. We now know that any particle big enough to be seen by the naked eye is far too large to show any movement due to atomic or molecular collisions and that the motion Democritus observed is due to thermal currents in the air. That Democritus gave a rational explanation but to the wrong phenomenon reflects no discredit whatsoever on his theory or his method. And the Brownian motion is still one of the most convincing pieces of evidence in support of the existence of atoms and their motion. (See experiments, p. 115, and, for a full discussion of this, see Sambursky, pp. 115–18.)

Atoms represent the permanence in the world and change is due to their constant and random motion and collisions (see Sambursky, p. 127). Since atoms can be regarded as points in motion, their movement is capable, in principle at least, of mathematical description; an ideal much prized by the Greeks and by scientists ever since. Another important consequence of this theory is that since matter consists of atoms differentiated only by their size, shape and motion, then all other qualities of matter must arise from the interaction of atoms with the sensory organs. Thus emerges for the first time the distinction between primary and secondary qualities, a distinction the analysis of which has proved to be a persistent and fruitful problem. Galen (A.D. 129–199) quotes Democritus on this distinction:

> Sweetness exists only by convention, bitterness by convention, colour by convention: the atoms and the void alone exist in reality. We have no accurate knowledge of anything in reality, but can be aware only of the changes which correspond to it in the conditions of our bodies.

Finally, this technique of the atomists of interpreting observable phenomena (evidence of the senses) in terms of hypothetical bodies (the products of reason) provided a most valuable method for the investigation of nature.

Like all good answers, atomism posed further stimulating questions. Perhaps the most important of these was how we are to regard atoms: are they real entities or convenient explanatory devices? This realist–phenomenalist problem is still with us and indeed arises acutely in contemporary matter theory.

PLATO

The mathematical possibilities of atomism were explored by Plato (429–347 B.C.), particularly in the *Timaeus* (see introduction and translation by H. D. P. Lee). Plato was much impressed by the certainty and precision of mathematics—first clearly demonstrated by the Pythagorean school—and believed that the uncertain and imprecise information which we receive through our senses should ideally be expressed in purely mathematical terms. Only then would our ideas be truly rational. (If truly rational in this sense, then truly abstract and completely removed from the world of experience. This can form the basis for a discussion of the compromise the scientist strikes between a purely rational and a purely sensory or empirical approach to nature.)

Plato suggested that the atoms of each of the four elements were in the form of a regular solid: fire—tetrahedron, earth—cube, air—octahedron, and water—icosahedron. But there are five—and, by the proof of Theatatus, only five—regular solids. For Plato, the fifth—the dodecahedron—represented most closely the cosmos, which, being perfect, was of the perfect shape—spherical, and the dodecahedron approximates most nearly to a sphere. (The Greeks made footballs by stitching together twelve pentagonal panels of leather.)

Plato, unlike Democritus, believed in transmutation—the conversion of one element into another. This followed as a geometrical consequence of his theory: the sides of a regular solid can be resolved into triangles and those put together again to form a different regular solid. (See *Architecture,* pp. 82–8, and the film *The God Within.*)

Now, although all this is very ingenious, it illustrates well the danger of over-mathematisation; it is too abstract and was devised with too little regard for the world it purported to explain. In consequence it helped little in the understanding of matter and it provoked few stimulating questions. But Plato's theory does have some significance: it was the only systematic attempt to combine the idea of an atom with that of an element and to explain reactions thereby until the work of Dalton at the beginning of the nineteenth century. Also Plato showed great wisdom in suggesting a limit to the possible number of different types (shapes) of atoms. Democritus' postulate that there are as many shapes of atoms as there are materials is vague and unhelpful. When some limit is imposed on the possible types of atoms conceivable, then the composition of matter and material change can be meaningfully investigated. Finally, though Plato allowed his use of mathematics to overreach the empirical knowledge of his time and thereby made his theory too fanciful and abstract, yet his method of starting from observable facts and devising a rigorous mathematical structure of hypothetical entities to account for them is one which from Newton onwards has proved to be most successful.

ARISTOTLE

Plato's greatest pupil Aristotle (384–322 B.C.) adopted a completely different approach to nature in general and to matter theory in particular. For Plato abstract mathematics was the ideal; for Aristotle, trained as a biologist (and whose reputation as such still remains high), it was the living world—the world of spontaneously interacting organised things, the world of birth, growth, decay and death—which represented reality and which must be explained first in any philosophy of nature. Aristotle was little impressed by mathematical abstraction; he emphasised and greatly developed the functional explanation of phenomena. He tried to explain all changes in terms of the organic processes (see *Architecture,* p. 94, and topic I.2, p. 32).

In the biological world things are interdependent; a creature is influenced by its environment and vice versa. Now this interdependence of things was a feature of nature largely ignored by the theories of Democritus and Plato. Atomism generally infers independence—two atoms, unless they are in contact by collision or combination, are separated by the void and can have no effect on each other. Because of this, Aristotle could not accept atomism; he had hit on its chief weakness—action at a distance, a weakness still of any 'pure' atomistic theory.

This problem of action at a distance was developed later by the Stoics. They taught that interdependence of bodies was achieved through an all pervading 'pneuma' (see p. 109). Action at a distance has proved to be one of the most intractable problems of matter theory; Descartes wrestled with it, Newton could not 'explain' gravitation (it was in this context that he said, 'I feign no hypotheses'), and today it is still present in the problems of particles in fields of influence. (For a very good account of the whole problem of action at a distance see *Forces and Fields* by Hesse.)

Aristotle objected to the void on purely philosophical grounds—'One cannot conceive of that which is not.' Nature's supposed abhorrence of a vacuum dates from Aristotle's refusal to accept any theory which involved the void.

The matter theory of Aristotle formed a part of his general cosmological theory. He believed that all materials arise out of the constant growth and decay (note the biological approach) of the four qualities we can distinguish: hot and cold, dry and moist. Appropriate combinations of these qualities give rise to the four elements of Empedocles: dry and hot give fire, moist and hot give air, dry and cold give earth, and moist and cold give water. These relationships are conveniently shown in the well-known diagram:

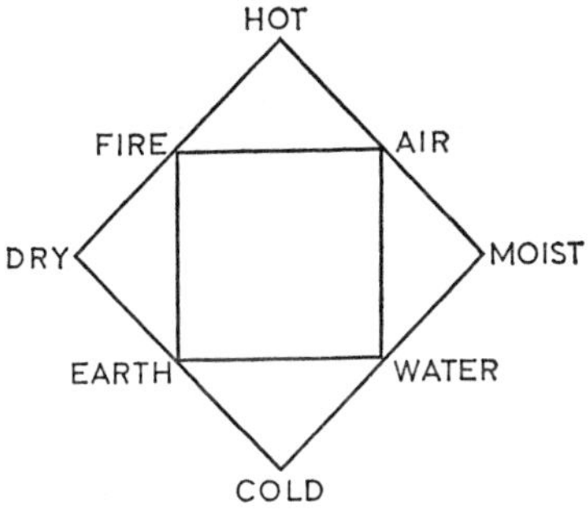

The most important feature of Aristotle's theory, and one out of which alchemy was to arise, was that just as the body is subject to constant metabolic changes, so too is (what we now call) the inorganic world. For Aristotle any piece of matter contained all four elements in a proportion which depended on the proportion of the four qualities. The air we breathe contains a high proportion of hot and moist and correspondingly less of the other qualities. Its properties therefore approach those of the 'ideal' element air. If, however, in the course of the general metabolism, the proportion of cold increases, then the air will turn into water (not a bad description of fog formation!). It is the qualities rather than the elements which are conserved and thus transmutation is possible, and indeed occurs in nature. If by art man can imitate nature, then he will bring about transmutation; this was the starting point of alchemy.

There was a fifth element in Aristotle's system, that of which the heavens are made. Here there are no changes; this does not mean that the heavens are lifeless but that everything there is perfect and since change occurs in order that perfection shall be achieved, the cosmos is changeless. The celestial fifth element is the quintessence, the essence of perfection (see Plato and the dodecahedron, p. 105). This contrast between the changing imperfect terrestrial world and the unchanging perfect celestial region was most important in astronomy (see Section I).

The idea that all changes occur in an attempt to attain perfection implies anticipation, a working towards some future event. (This teleological approach helps to account for the ready assimilation of Aristotelian philosophy into Christian doctrine, where present good works lead to future blessedness.) Scientists subsequently found it better if changes are explained in terms of what has gone before. But the problem (yet another fundamental problem) of cause and effect has troubled philosophers since Greek times and is still with us.

The hitherto successful present-arising-out-of-past view of cause and effect is proving difficult when applied to certain events encountered in contemporary physics, and some have suggested that a teleological type of explanation may be more satisfactory (see topic I.2, p. 32).

It is worth asking why this biological approach to Aristotle was rejected by the scientists of the revolution in the seventeenth century. After all, Aristotle did obey the first commandment of science—build on what you see. We see birth, growth and death in a way that we do not see atoms, or forces, or frictionless pulleys, or ideal gases. Looking back we can say that only when the relatively simple principles underlying inanimate nature were properly understood could any sense be made of the enormously complex situations associated with living things. The emphasis on physical rather than biological science led to the discovery of relatively *few* fundamental theories which could account satisfactorily for the *many* phenomena (the appeal of economy). It is difficult to see how this could have happened if men had continued with Aristotle's biological approach. But some acute contemporary problems arise from our having pursued the divisive philosophy too far; sub-atomic particles, atoms, molecules, matter, stars and men are not completely independent and perhaps Aristotle's emphasis on the interdependence of things was not so wrong after all.

THE STOICS

In order to account for independence, the Stoics (found by Zeno *c.* 300 B.C.) preached the principle of *continuum,* which is in many respects the opposite of atomism since it denies the existence of entirely independent particles. According to the Stoics, all parts of the universe were connected by an all-pervading, warm and living breath or 'pneuma'. By means of the concept of pneuma (originally derived from Anaximenes and Heraclitus), they were able to face the problem of interdependence with more precision than was Aristotle. This approach to matter is somewhat akin, in principle, to the all-pervading wave patterns of wave mechanics, devised in this century to explain phenomena which were generally ignored by the nineteenth century with its almost exclusive belief in atomism. (See *Architecture,* p. 108, and *Quanta and Reality.*)

EXPERIMENTS AND DEMONSTRATIONS

(1) The principle behind Empedocles' 'water-clock' experiment (p. 102) may be demonstrated quite simply with a glass tube of about one-half inch diameter immersed in a 'gas-jar' of water: if the tube is dipped into the water while the thumb blocks the top end, the water does not enter; when the thumb is released, the water rises up the tube and the expelled air will blow out a lighted match.

(2) The random motion of atoms can be imitated by means of polystyrene balls moving in a wide glass tube; the balls are activated by compressed air blowing up through a zinc mesh base.

(3) The Brownian motion can be demonstrated by mixing six drops of Stephens' drawing ink with 5 ml of water. Put one drop of this mixture on a slide, gently lower on to it a cover slip and view under a magnification of about 500×. Alternatively, artists' 'gamboge' in water can be projected using a $\frac{1}{6}$-inch objective.

(4) There are many good examples illustrating the fallibility of the senses in Abercrombie, *The Anatomy of Judgement.*

(5) Suppliers of mathematical models provide models of the five regular solids.

II. 3 Alchemy

A BRIEF HISTORICAL SURVEY

The period of alchemy extends from about 300 B.C. to A.D. 1600. It is impossible to deal with such a period in detail, nor is this necessary; the aim here is to give some idea of what alchemy was about and thereby to show that it was a logical consequence of Greek matter theory and that it paved the way for the more rapid developments in the seventeenth century.

The history of matter theory during this period follows quite closely that of Greek learning in general. After about 200 B.C. there was a general decline in learning in Greece. The ideas of the great Greek thinkers spread throughout the Mediterranean world and found a new focal point at Alexandria. (See map I.1, p. 16, also *Architecture,* p. 121, and Stillman, ch. 4.)

Alexandria was founded in 332 B.C. and rapidly became the greatest port and centre of trade in the Mediterranean. As in the case of Miletus the cosmopolitan vigour and business atmosphere of such a centre made it an important cultural 'melting pot' wherein ideas from any sources became intermixed and wherefrom many new ideas grew. In this atmosphere the great museum and library fulfilled the intention of their founder and formed the centre of Greek influence.

The traditions which came into contact and conflict and which influenced alchemy were: from the Egyptians, the by now highly developed craft tradition wherein man showed a wide knowledge of materials and great skill in their use; from the Greeks, the spirit of rational enquiry and in particular from Aristotle, a vast, logically coherent cosmological system; and from the Persians, the Jews and (later) the Christians, elaborate mystical religious thought.

In A.D. 641 Alexandria was sacked by the Arabs. The Greek and Hellenistic manuscripts they found were removed to towns such as Baghdad. Here they were preserved and translated. In the case of those manuscripts concerned with astronomy and alchemy the Arabs incorporated ideas and discoveries of their own.

From about 1100 onwards European scholars, such as Gerard of Cremona and Robert of Chester, travelled in the middle east, often at great personal risk, seeking out the early writings and translating them into Latin. Thus the bulk of the writings of the Greek philosophers passed into Western medieval culture (see Crombie, *Augustine to Galileo*).

From about 1200, centres of learning appeared in Europe at Paris, Bologna, and later at Oxford and Cambridge. These were part of an intellectual revival which gained momentum as a result of the recovery of the Greek heritage from the East. Thus dawned the period of medieval scholasticism, a period when, under the powerful hand of the Church of Rome, the minds of men were chiefly concerned with reconciliation of Greek philosophy to Christian dogma, and with the preparation for the life hereafter.

Alchemy, and particularly later alchemy, reflects fairly accurately the characteristics of medieval scholasticism: a reliance on the writings of the Greeks, especially Aristotle; a preoccupation with the interpretation of these writings and an unwillingness either to entertain new ideas or to look at old ideas in new ways suggested by experience; the use of unnecessarily complex language and, in general, a tendency to be obscure rather than clear and dogmatic rather than critical.

THE REASONABLENESS OF ALCHEMY

One of the chief virtues of devoting a short time to alchemy in a course on matter theory is that it provides a useful exercise in 'mental transposition'. In the light of modern knowledge and the history of its recent development, it is too easy to dismiss alchemy as worse than useless, a positive hindrance to the development of chemistry. But if we put ourselves into the intellectual situation of, say, 1300 we shall see that alchemy was reasonable: it was built on Greek matter theory and had developed a logical and coherent structure. Furthermore it was, initially at least, directed towards a high ideal—the perfecting of matter. That base metals could be fairly easily turned into alloys which could be passed off as gold, and that such techniques could be the source of easy money, was not alchemy but pseudoalchemy. Alchemy should not be confused with, nor condemned for, such practices, just as science must not be confused with nor condemned for the pseudoscience that litters much contemporary advertising. The misuse of the elaborate jargon and ideas of an honest and worthwhile study in order to dupe unsuspecting and uncritical customers is a practice that was by no means confined to the Middle Ages! When in the sixteenth and seventeenth centuries alchemy declined (its chief fault was that it lived too long), new and intellectually more profitable approaches to matter arose from it. The 'iatrochemisty' (medicinal chemistry) of Paracelsus and Van Helmont and the 'modern' chemistry of Boyle relied heavily on the alchemists for their techniques, their terminology and their inspiration.

Alchemy represents an attempt to combine Greek philosophy with the craft traditions of the middle-eastern civilisations. For those nurtured in the business-like atmosphere of Alexandria, Greek thought was too abstract and speculative. Though it gave undoubted intellectual satisfaction by its general plausibility, when applied to specific instances it was found wanting. The rather fragmentary observations of nature which provided the only evidence for Greek ideas were too general to withstand close scrutiny. On the other hand the craft tradition afforded an abundance of observations which arose from an active and close interrogation of nature. But in this tradition there was not the intellectual stimulus which comes from an explanatory theory devised by standing back from the canvas of nature. In the absence of such theory progress is slow for want of drive and direction.

The Alexandrian alchemists remedied the shortcomings of these inherited traditions by attempting to combine them. The result might seem a failure in the light of modern science, but that the alchemists were the first to make the effort gives them an important place in the history of chemistry. Indeed, it is difficult to see how Boyle could have set chemistry on what we *now* see was the right path, had his critical faculties not been sharpened by theories of the alchemists and had he not inherited the enormous wealth of their experimental observations.

THE DEVELOPMENT OF ARISTOTLE'S THEORY

Alchemy was based on Aristotle's theory of matter (see p. 107). This was concerned with the proportion of the four qualities: hot, dry, cold and moist. By changing the proportion of these, the proportion of the four corresponding elements—earth, fire, air and water—was changed and thus one form of matter was changed into another. For Aristotle and the alchemists 'matter' was that which changed. The attention of the alchemist was focussed on metals, for gold was reckoned to be perfect matter, and thus by turning base metal into gold the alchemist was assisting the divine purpose and furthering his own spiritual well-being. This religious attitude towards work shows the influence of oriental and early Christian writers; men came to regard intellectual insight not as an end in itself, but as a means to gaining spiritual salvation.

It is important to realise that the sharp distinction we now draw between animate and inanimate matter did not then exist. Things are like ourselves; we recognise (but still do not understand) a spiritual part which controls the bodily part. This idea of controller and controlled was applied to ordinary matter.

Thus the problem of transmutation becomes that of creating conditions under which the maturing or perfecting will occur. How are we to know these conditions? To transpose our mind into that of an alchemist is no easy matter, but to try to imagine the feelings of a lump of adolescent copper and its aspirations to become gold is even more difficult! This was the intellectual task the alchemist set himself. (Chemical anthropomorphism is still far from dead—one is sometimes tempted to think, and some textbooks actually say, that because of their electronic structures sodium and chlorine *desire* to combine to form a stable ionic bond. And much of the qualitative description of such things as shifting equilibria, the formation of hybrid bonds and chemical thermodynamics smacks of anticipatory anthropomorphism.) The technique used by the alchemist to gain this insight into the inanimate world was most sensible and obvious: imitate nature.

According to Aristotle all changes are, in principle, possible, though some do not occur directly. An egg does not turn directly into a hen but *under certain conditions* it will pass through several phases (Aristotle studied closely the development of the embryo in the egg) until finally it becomes the fully-grown hen. This is a move towards perfection. Now, copper does not change directly into gold. How is the change to be achieved? What are the necessary conditions? From observing the chick we can see that a seed is necessary and so also is warmth, so we add to our copper a 'seed' of gold and warm the mixture, preferably in a vessel in which we shall be able to follow the progress of the copper. Thus we can appreciate the importance of the furnace and the glass retort in the alchemist's laboratory. Since the alchemist did not distinguish between living and non-living things as we now do, he believed that the process of growth goes on in the earth, not only for plants but for metal ores also, though the growth of the latter was at a slower rate. Thus in the 'womb of the earth' the ores of the base metals are, under divine influence, turning into gold. The retort on the furnace was merely an attempt to speed up the process. (Sometimes nature helped in the deception: when galena is heated in air for a long time, the lead is apparently turned into genuine silver. The ninety-five per cent of lead in the mineral is oxidised by the air, leaving the five per cent of silver the mineral contains.)

SPIRITUALISM IN ALCHEMY

The crucial problem for the alchemist was how to induce the divine influence into his operations. He believed, following Aristotle and the Stoics, that the change in the form of matter—corresponding to the change in the proportion of the four elements—was caused by the 'spirit' or pneuma. It was chiefly through the pneuma[1] or breath of life that divine influence operated. The activity of the spirit in the development of minerals and metals, like that of crops, was governed by celestial changes. This accounts for the carly association between the common metals and the planets (see *Architecture,* p. 129), and for the close connection between alchemy and astrology.

The concern with spiritual matters had certain important practical consequences. It accounted for the great attention given to the process of distillation—whereby the spirit was separated from the body *(caput mortuum)*. Whatever the motives of the alchemist for carrying out distillations they did lead to the isolation of many important substances, e.g. alcohol, sulphuric acid and nitric acid. Interest in spirits and vapours led the later alchemists to accord particular importance to sulphur and mercury.

According to the sulphur-mercury theory, metals differed chiefly in the proportion of these two vapours; sulphur was the dry vapour and mercury the moist vapour. Transmutation was effected by changing the proportion of these two substances. Though this may now seem worthless and far-fetched, there are

[1] The difficulty we now have in portraying the precise meaning of this word is due to the disappearance of the idea from our system of thought.

certain observations which, in the lack of any precise ideas on chemical constitution, would lend support to this theory. For example, the winning of metals from sulphidic ores is accompanied by the evolution of sulphurous fumes (of sulphur dioxide); some sulphidic ores have a metallic lustre ('fool's gold'); and the amalgamation of zinc by rubbing with mercury leaves a silver-like surface. (See experiments below.)

An interesting and sophisticated idea in later alchemy was that of the 'elixir' or philosopher's stone. This represents a development of the 'seeding' principle in that it was thought that the elixir would bring about the change of base metals into gold without itself being consumed in the process. Thus the function of the elixir very closely resembled that of the catalyst in chemistry. Since it was believed that the philosopher's stone would bring about perfection in matter, and since the alchemist regarded himself and his experiment as a spiritual whole, it was reasonable to hope that the stone would remove the corruption from man and bestow immortality. Perhaps as a final tribute to the much-maligned alchemist it is worth pointing out that though few scientists would now claim a spiritual oneness with their experiments, in recent research it has become increasingly difficult for the scientist to preserve that complete independence of the experiment from the experimenter which was assumed by the classical scientists of the last three centuries.

EXPERIMENTS AND DEMONSTRATIONS

(1) Many of the experimental techniques first introduced into chemistry by the alchemist can be demonstrated: sublimation, crystallisation, filtration, and distillation; it is interesting to compare distillation apparatus illustrated in alchemical works (and reproduced in many of the books in the bibliography) with standard arrangements now used for straight, reflex and extraction distillation.

(2) Some simple 'pseudo-alchemical transmutations' may easily be demonstrated. Clean a piece of zinc in dilute sulphuric acid then rub it gently with mercury—a silver-like layer of zinc amalgam will be formed. Suspend a strip of copper over a small piece of arsenic in a test tube. Heat the tube gently (in a fume cupboard—the fumes are poisonous); the copper will slowly become coated with a silver-like layer of copper arsenide.

II. 4 'Element' and 'atom' in the seventeenth century

INTRODUCTION

The most important feature of matter theory in the seventeenth century was the general sense of *awakening*. Awakening to precisely what was not clear even by 1700, though by that time the scientific revolution in astronomy was almost complete. In matter theory there were only a few promising strands of thought which were not drawn together into a coherent theory until the nineteenth century.

There is a problem in the selecting of material with which to illustrate this awakening in the seventeenth century. The period is sufficiently near for there to be an abundance of writings fairly readily available; yet much of this material is so obscure that it is fit only for the specialist. Any attempt to clarify the issues by presenting them in a simple way gives a false impression of coherence and precision. In order to demonstrate this it is a good idea to read passages from, for example, *The Sceptical Chymist* or the queries in *Opticks* other than those quoted here; the confusion and inconsistency will be apparent. This must be borne in mind when reading what follows, where the need for brevity and clarity leads to an incomplete and perhaps too coherent a picture.

Such a false picture suggests too sharp a break between alchemy and chemistry; the latter grew slowly out of the former. It is true that Boyle seriously undermined the fundamental ideas of alchemy but it must be remembered that, as was emphasised in topic II.3, he owed much to the alchemical tradition and also that much of this tradition persisted throughout the seventeenth century and into the eighteenth century until it was expelled by the quantification of chemistry.

Perhaps the best way to present the significance in the seventeenth century is to consider the development of the ideas associated with the terms *element* and *atom*. These terms originated, as we have seen in topic II.2, in Greek science, and in topics II.6 and II.7 is discussed their significance in the coherent theories of the nineteenth century. It would not be too great an over-simplification to say that the history of matter theory is essentially the meanings which have been attached to these terms over the last two and a half thousand years. In the seventeenth century much of the earlier confusion was removed and the minds of men were prepared for the more precise and helpful meanings the terms subsequently assumed.

ELEMENT

Though there was general agreement on the definition of elements —those substances into which all other materials could be resolved —there were conflicting opinions as to the substances which qualified as elements.

Peripatetics, spagyrists and Paracelsus

The peripatetics were those who followed Aristotle and believed in the four elements of Greek science—earth, fire, air and water. Others followed the tradition of the later alchemists and believed that matter was made up of sulphur, salt and mercury. Those who followed this tradition were called 'spagyrists' and the most influential amongst these was Paracelsus:

> Sulphur represents the gaseous and combustible elements, also the forces of the soul, or the principle of energy; Mercury represents the fluid elements, also the forces of the intellect; Salt stands for the solid elements, also for the principle of matter. *(Opus paramirum)*

> No one shall say that metals are dead matter. On the contrary, their salt, sulphur and quintessence have a great power to activate human life. *(Archidoxa)*

These are taken from Pachter, and illustrate the confused state of science. Pachter admirably sums up Paracelsus:

> Somewhat helplessly he struggled with notions approaching it [element] but *did not know how to express his ideas*. It remained a 'great mystery' how chemical substances seemed to live, to react on others, to attack and to preserve themselves. But the action was there, undeniably, and it had to be used and explained.

We have italicised words which might form the basis for a discussion on the problem of the communication of ideas which are 'ahead of their time'. Also, incidentally, the quotation from *Archidoxa* might remind some student of the supposed curative properties of copper bracelets currently in vogue.

Around 1600, scientists were not sharply divided into 'peripatetics' and 'spagyrists': many held a composite philosophy of the four Greek elements on which were superimposed the three alchemical principles.

Van Helmont

Van Helmont, the greatest of the medical chemists, believed water was the fundamental stuff from which materials were formed. He believed he proved it by his celebrated 'willow tree experiment' (see *Architecture,* p. 169). This experiment is worthy of emphasis as being the first clearly described quantitative experiment. Van Helmont had a clear aim in view. He took all reasonable steps to ensure the quantitative accuracy of the experiment, and his interpretation of the results is perfectly sound *within the framework of his knowledge.* (Note that in his interpretation he *assumes* conservation of mass during the reaction.)

It should be easy for the scientists in the class to expose the fallacy in the interpretation and give the modern explanation. The irony of the situation was that van Helmont's error was failure to take into account the part played by gases of the air and carbon dioxide in particular. He was the first to distinguish different gases and of all the gases he describes carbon dioxide most clearly. The term 'gas' was first introduced by van Helmont and referred to carbon dioxide. Though he described qualitatively gases which we now recognise as different chemical substances, van Helmont never used the plural 'gases' and seems to have regarded them as manifestations of a spirit *in* matter rather than states *of* matter. He had some unfortunate explosions with gases under pressure and concluded gases could not be collected in vessels. The significance of van Helmont in matter theory is difficult to assess. His work on gases was a step forward but his influence on others, especially Boyle, was probably more important than his own ideas. Though there is much sound sense in his teaching, his writing is obscure (see *Architecture,* pp. 166–73).

Boyle

There has been some misunderstanding of Boyle's ideas on elements. *The Sceptical Chymist* (1661) is a detailed account of evidence against the previous 'element' theories—the four Greek elements and the three Paracelsan principles. In the course of his refutation of these theories, Boyle gives his criteria by which elements should be defined (see *Architecture,* p. 204). This definition is often quoted alone and thereby Boyle is extolled as the founder of the modern theory of elements. There are two objections to this: the definition he gave was not new but merely an amplification of the idea of an element first clearly expressed by Empedocles (see p. 101), an idea which lay behind the Greek elements and the Paracelsan principles as much as it does the 'modern' element. Secondly, we read in *The Sceptical Chymist* following the above definition:

Now whether there be any one such body to be constantly met with in all, and each of those that are said to be Elemented bodies, *is a thing I now question.* (Our italics.)

Boyle's superiority over his predecessors is shown not in any radically new idea of an element, but in recognising that *on the available experimental evidence* there was little justification for defining *any* substance as an element. He was confirmed in this view by extensive experimental researches which led him to the conclusion that matter was infinitely transformable by chemical means (see *Architecture,* p. 203). Such a view is the opposite to the modern idea that matter can only be transformed in limited and specifiable ways, an idea in which the notion of an element forms an obvious part.

Though Boyle's experiments did not lead him to the elements, they did much to prepare the ground for the fruitful revival of the idea a hundred years later. He undermined many of the older views on reactions. For example, he rejected the contention that analysis by fire produced the same elements from all substances. He showed that supposed elements like spirits and acids were heterogeneous and not immutable. (In support of this he claimed to have converted water into earth—a claim which Lavoisier showed to be fallacious.) He showed that salt, 'the element of solidity', was not necessary for precipitation by producing precipitates in the absence of salt. Likewise he proved wrong the idea that all reactions involve an acid–alkali combination, by demonstrating reactive substances which could not be classed as acid or alkali. (For other examples, see Stillman, ch. 10.)

In the many experimental observations of this kind, Boyle's aim was not only to undermine previous theories but to provide a more reliable foundation for subsequent ideas. He realised the need for far greater care in classifying chemical substances and for a reliable system of analysis. His work went a long way towards providing these needs (see *Architecture,* p. 206). To illustrate this it is interesting to compare two accounts of similar chemical reactions. The first is by Nicolas Lemery, the distinguished French chemist of the 'old school' and a contemporary of Boyle:

When one throws corrosive sublimate into lime water, it takes on a yellow colour and loses so much of its corrosive force that one can safely take it by mouth. I do not aspire to explain this change of colour; I prefer to leave it to those with more leisure than I to explain the disposition of the pores that the acid and the lime must give the mercury so that it may modify or reflect the light in such a way that a material which formerly appeared very white is now yellow. I shall only say that the lime water softens or weakens the force of the sublimate because of the particles of lime which it contains, which particles, meeting and striking the sublimate, break part of the points in which the corrosiveness consists.

Boyle wrote that this reaction was very like those 'of metalline bodies corroded by acid salts; so that the colour in our case results from the coalition of the mercurial particles with the saline ones, wherewith they were formerly associated, and with the alkalisate particles of the salt of tartar that swim up and down in the oil'.

That is to say, when mercury is joined to saline particles, the resulting compound is white; when it is joined to 'alkalisate' particles, the resulting compound is yellow. This makes far more sense in chemical terms than does Lemery's explanation, and is, besides, much more likely to lead to profitable chemical investigations than any discussion in terms of purely physical principles (like Lemery's particles) or purely chemical principles (salts, oils, mercuries). (Examples and comment taken from Boas' paper in *Critical Problems in the History of Science.*)

Boyle was apparently much more of a 'modern' in his attempts to interpret chemical reactions in terms of atoms. Although, as we shall see later, one must be careful even in this connection not to attribute to Boyle too modern an attitude, his approach to matter was characteristic of the seventeenth century and represents a significant improvement in matter theory.

THE REVIVAL OF ATOMISM

Greek atomism never died. After Aristotle and particularly after the Stoics, atomism sank into the background of scientific thought but there is a continuous, if at some times tenuous, 'atomic' tradition. Gassendi attempted to remove philosophical (and religious) objections to atomism and thereby made it more acceptable to seventeenth-century thinkers. (See Harré, *Mind and Method,* p. 63.)

Descartes' influence was more indirect but no less considerable. He accepted a corpuscular view of matter but could not accept the void. Space therefore is filled with particles of different fineness. (For an account of Descartes' theory see *Architecture,* p. 173, and Hall, *The Scientific Revolution,* pp. 207–13). In spite of the strange 'compromise' nature of Descartes' matter theory, its mechanistic nature appealed to many of the 'pure' atomists and Descartes' analytical geometry was of great use in the development of mathematical atomism. (For the importance of 'mechanism' see *Architecture,* p. 183.)

Galileo adopted atomism in the hope of bringing matter theory within the fold of mathematics. (This has always been for some a powerful reason for accepting atomism—see p. 104.)

By means of an atomic theory, Galileo was able to differentiate 'primary' and 'secondary' qualities of matter (see p. 104). Primary qualities are those which are mathematically analysable such as weight and size. Other properties, such as colour, smell, warmth, etc., merely arise from our contact with the body and are therefore not essentially part of the body; they are secondary qualities. A nice illustration of Galileo's reasoning is in *Il Saggiatore* (1623):

> An example, I believe, will clearly explain my concept. Suppose I pass my hand, first over a marble statue, then over a living man. So far as the hand, considered in itself, is concerned, it will act in an identical way upon each of these objects; that is, the primary qualities of motion and contact will similarly affect the two objects, and we would use identical language to describe this in each case. But the living body, which I subject to this experiment, will feel itself affected in various ways, depending upon the part of the body I happen to touch; for

example, should it be touched on the sole of the foot or the kneecap, or under the armpit, it will feel in addition to simple contact, a further affection to which we have given a special name: we call it 'tickling'. This latter affection is altogether our own, and is not at all a property of the hand itself. And it seems to me that he would be gravely in error who would assert that the hand, in addition to movement and contact, intrinsically possesses another and different faculty which we might call the 'tickling faculty', as though tickling were a resident property of the hand per se. Again, a piece of paper or a feather, when gently rubbed over any part of our body whatsoever, will in itself act everywhere in an identical way; it will, namely, move and contact. But we, should we be touched between the eyes, on the tip of the nose, or under the nostrils, will feel an almost intolerable titillation—while if touched in other places, we will scarcely feel anything at all. Now this titillation is completely ours and not the feather's, so that if the living, sensing body were removed, nothing would remain of the titillation but an empty name. And I believe that many other qualities, such as taste, odour, colour, and so on, often predicated of natural bodies, have a similar and no greater existence than this. (Quoted in *Philosophy of Science* by Danto and Morgenbesser.)

The chief scientific objection to atomism had been the necessity for the void. Much of the weight of this objection was removed as a result of the work of Torricelli, Viviani, Von Guericke, Pascal and Boyle on atmospheric pressure, air pumps and barometers. This work is most valuable for illustrating the methods of scientists. (For a very full account and discussion see Conant, *On Understanding Science,* ch. 2 or *Harvard Case Histories,* vol. I.)

BOYLE AND ATOMISM

These various ideas about atomism all influenced Boyle in his attempts to apply atomism to physical and chemical problems, and it is by looking at his work that we get probably the clearest idea of the success and failure of atomism in the seventeenth century.

Although Boyle was unable to give a clear or systematic account of chemical reactions in terms of corpuscles, he did show that explanations in such terms were much more fruitful than those involving vague 'principles' and 'elements'. Boyle's chief aim was to show that *qualitative* changes, such as chemical reactions, are due to *quantitative* changes in the minute bodies which make up substances. He did not believe in any forces of attraction between reacting atoms; chemical reaction occurred when the size and shape of the particles of one substance corresponded to the size and shape of the pores between the particles of another substance. He shows that such diverse changes as the growth of plants, the development of the chick in the egg, the growth of crystals from solution, as well as more familiar chemical reactions involving changes in crystalline form, colour, texture, and taste are better explained in terms of the rearrangement of constituent particles than by the older ideas of substantial forms, etc.

He concludes:

For in the experiments we are speaking of, it cannot well be pretended, or at least not well proved, that any substantial forms are the causes of the effects I have recited; for in most of the (above-mentioned) cases, besides that in the bodies we employed, the seminal virtues, if they had any before, may be supposed to have been destroyed by the fire, they were such, as those I argue with would account to be factitious bodies artificially produced by chemical operations. And it is not more manifest that in the production of these effects there intervenes a local motion and change of texture by these operations, than it is evident and precarious that they are the effects of such things as the schools fancy substantial forms to be: since it is these new experiments, by the addition of some new particles of matter, or the recess or expulsion of some pre-existent ones, or, which is the most frequent way, by the transposition of minute parts, yet without quite excluding the other two, that no more skilful chemist than I have been able to produce by art a not inconsiderable number of such changes of qualities, that more notable ones are not presented as by nature, when she is presumed to work by the help of substantial forms: I see not why it may not be thought probable that the same catholick and fertile principle, motion, bulk, shape, and texture of the minute parts of matter, may under the guidance of nature . . . suffice likewise to produce those other qualities of natural bodies, of which we have not given particular instances. (From *Origin of Forms and Qualities,* quoted in Harré, *Matter and Method,* p. 84.)

He did not, of course, get very far in this but he at least redirected the endeavours of chemists along more fruitful paths. Boyle was prevented from making much progress by his belief that the number of possible products in a reaction was unlimited (see *Architecture,* p. 203, for example. Because of this, Boyle's atomism, although seemingly modern in some ways (e.g. the idea of molecules, *Architecture,* p. 202), could hardly be said to anticipate the nineteenth century.

Hooke

It is worthwhile drawing attention to the significance of Hooke's *Micrographia* (1665—now available in facsimile by Dover). In this he discusses the corpuscular philosophy connection with crystal shapes (p. 82). Perhaps less specific but more important was the influence of the whole of *Micrographia,* for Hooke showed that the smaller the body often the more complex is its structure. This helped to prepare minds for belief in atomism. Any opportunity to show this book should be seized upon, if only for the magnificent illustrations.

Newton

It is right and proper that Newton's monumental contributions to physical science should be considered superior to his work in chemistry but it is a pity that as a chemist his reputation is sometimes little better than that of a dark dabbler in alchemy. Out of his interest in alchemy grew a penetrating and perceptive approach to chemistry. This approach to matter theory was very similar to that of Boyle (with whom he frequently corresponded). He regarded chemistry as a physical science rather than a mystical art, and adopted and refined the 'corpuscular' theory. Perhaps his most important contribution to the theory is the idea of inter-atomic attraction (see *Architecture,* p. 212). Unlike Boyle, Newton did believe in forces between bodies and suggested the existence between reacting atoms of a 'sociability'. This was later amplified into a fairly elaborate theory of selective attractions. A good example of this principle is to be found in his definition of acids—'substances endued with a great attractive force, in which force their activity consists'. This led to a crude but helpful table of affinity. In *De Natura Hudorum* (1692) Newton also suggests that particles associate to form aggregates 'of the first composition', that these associate to form aggregates 'of the second composition', and so on. He thus distinguishes transmutation from chemical reaction. (See *Architecture,* p. 221; for Newton and astrology, see topic I.6, p. 68.)

One of the richest sources of information on Newton's thoughts about matter and forces of attraction in particular are the 'Queries' at the end of *Opticks*. Here Newton allows his genius to range over his whole scientific experience.

After many pages of brilliantly reasoned speculation he comes to his general hypothesis:

All these things being consider'd, it seems probable to me that God in the Beginning form'd Matter in solid, massy, hard, impenetrable, moveable Particles, of such sizes and Figures, and with such other Properties, and in such Proportion of Space, as most conduced to the End for which he form'd them; and these primitive Particles being Solids, are incomparably harder than any porous Bodies compounded of them; even so very hard, as never to wear or break in pieces; no ordinary Power being able to divide what God Himself made one in the first Creation.

(For several examples from 'Queries' see *Architecture*, pp. 211–15.)

During the second half of the seventeenth century several attempts were made to estimate the size of constituent corpuscles. Newton derived a value from his rings in thin films (see *Architecture*, p. 216). His friend Edmund Halley devised an ingenious method of estimating the size of an atom of gold. He had a gold beater beat out a lump of gold of known volume into a sheet as thin as possible (i.e. to maximum area). Then dividing the volume by the maximum area gives the thickness of the sheet which, Halley assumed, corresponded to one atom. The result was inaccurate but the method is quite sound. Modern methods of determining molecular size by the thickness of thin oil films are essentially the same.

II. 5 The problem of combustion

In the last topic we have seen the beginnings of modern chemistry, though in 1700 there was still much confusion. Even Boyle with his often brilliant insight into chemical matters could do little more than indicate in a rather general way how this confusion might be resolved. Looking back it seems that perhaps the most serious handicap to progress was the too catholic outlook of chemists: 'all nature their province' was at once an imposing yet inhibiting manifesto (cf. the Greeks, Boyle tackled problems from every part of chemistry and in every part there was contradiction and confusion. *It so happened* (and one cannot be more explicit than this) that in the eighteenth century chemists became concerned chiefly with a relatively small but important part of chemistry—that concerned with combustion and calcination. Thus blinkered, they discovered relatively clear and unambiguous explanations for a few important chemical reactions. From these, chemists of the nineteenth century were able to extend the understanding of chemistry as a whole.

Although these problems of combustion and related reactions were relatively few, to follow even a simplified account of the progress in the eighteenth century is difficult. In presenting this topic it is a good idea to demonstrate first in outline, and if possible experimentally, the chemistry involved. A list of simple experiments is given on page 133. If this policy is adopted it should be made clear that combustion, calcination and respiration are chemically similar since they all involve the combination of a substance with oxygen. (Though respiration should be mentioned, any detailed discussion is best omitted in a short course.) The products in these processes are oxides; some oxides are gaseous and if the reaction is carried out in an open vessel they 'disappear' into the atmosphere; the majority of the oxides are solid, so when the oxygen combines there is an increase in weight. If the reaction is carried out in a closed vessel there is no change in weight; the oxygen is merely transferred from the atmosphere to the oxide. Oxygen occupies about a fifth of the atmosphere, so after combustion in a closed vessel, about a fifth of the air is consumed.

If these reactions are carried out in pure oxygen, they are more vigorous and the whole of the enclosed gas can be consumed. Mercury is unusual in that on gentle heating in air or oxygen it forms mercury oxide, which on stronger heating splits up again into mercury and oxygen. Finally, carbon and hydrogen when heated with most metal oxides combine with the oxygen leaving the metal.

The historical discussion could begin with a brief reference to the ideas of Ray, Hooke and Boyle on combustion. Both Ray and Hooke gave speculative accounts of combustion which agree quite closely with the modern view. In *Micrographia* (1665—see Dover facsimile edition, p. 105) Hooke says:

> Flame seems to be a mixture of Air, and the combustible volatile parts of any body, which parts the encompassing Air does dissolve or work upon, which action, as it does intend the heat of the aerial parts of the dissolvent, so does it thereby further rarifie those parts that are acting, or that are very near them, whereby they growing much lighter than the heavier parts of that Menstruum that are more remote, are thereby protruded and driven upward.

Compared to the phlogiston theory these statements seem 'modern' in that Hooke (and Ray, whose ideas were similar —see *Architecture,* p. 229) suggested that combustion involves interaction with, and removal of something from, the air. It is important to emphasise that such statements, though interesting in retrospect, do not constitute a satisfactory theory of combustion. They are too speculative and too general; the phlogiston theory, which we now say was 'wrong', did give a reasonable and precise explanation of a variety of specific chemical reactions and was therefore, *at the time,* a 'better' theory. Boyle's experiments on calcination and his interpretation of them (see *Architecture,* pp. 228–9) provide interesting material for a discussion of chance in scientific discovery: 'If he had weighed the flask before breaking the seal—what then?' A point omitted from *Architecture* is that on breaking the seal Boyle heard the whistling of the air entering the flask, but ignored it. Notice in *Architecture* the point that Boyle *could not ask* the questions which Lavoisier later asked and answered. He was asking the question: 'Can fire be weighed?' not 'What is the role of air in combustion?' This probably explains why he was 'blind' to what *to us* seem significant features of his experiments.

THE PHLOGISTON THEORY

This theory is first found in the writings of Becher. In his *Physicae Subterranea* (1669) we can see that the fundamental assumption of phlogiston theory—that something is *expelled* from a burning body—is a logical consequence of alchemical theory of which Becher was a follower. His matter theory was essentially a refinement of the three principles of Paracelsus. For the Paracelsan 'sulphur' Becher substituted 'oily earth'—*terra pinguis*. It was assumed by the alchemists (following Aristotle) that the combustion of a substance always led to a breaking-down into simpler constituents. It was for this reason that Becher assumed that a simple body could not burn. If a body was combustible it must contain *terra pinguis* which was expelled during combustion.

Becher's ideas were clarified and improved by his pupil Stahl and it was Stahl who introduced the term 'phlogiston' for the inflammable principle in combustible matter. From his writings we can give an outline of the theory:

(a) Phlogiston is the principle of inflammability. A combustible body contains phlogiston which is lost on burning.

(b) Phlogiston is imperceptible to the senses when uncombined. It can never be destroyed.

(c) Substances such as charcoal, oily and fatty substances (hydrogen was added later) are especially rich in phlogiston.

(d) Incombustible substances, such as calces (oxides in modern terms) of metals, have either never possessed phlogiston or have lost it.

(e) Certain calces may be converted into the metal by heating with phlogiston—rich substances such as charcoal:

$$\text{Metal} = \text{calx} + \text{phlogiston}$$

(f) For the production of flame both phlogiston and free air are necessary.

In the writings of later phlogistonists the function of the air was regarded as an absorber of phlogiston; things ceased to burn in an enclosed volume because the air was saturated with phlogiston.

The fact that there was often an increase in weight accompanying a *loss* in phlogiston was accounted for either by the particles becoming more compact, giving an increase in density, or by heavy air particles becoming lodged in the calx. This problem of the increase in weight on calcination was a constant source of trouble and ultimately contributed to the downfall of the theory. In the early part of the eighteenth century, however, the terms mass and density were imperfectly understood, so the explanations offered were reasonable.

At this stage it is important to emphasise that until the problems associated with gases were solved (see below) and until ideas on density, mass and its conservation were clarified and accepted, the phlogiston theory was perfectly adequate. It was logical and it combined into a single scheme many otherwise unrelated facts. It also posed fruitful problems such as those on the nature of gases and on quantitative changes in chemical reaction. (See *Architecture* (pp. 237–41) and White. Conant in *On Understanding Science* (p. 81) and in *Harvard Case Histories* (vol. 1, ch. 2), gives a very good discussion of the phlogiston theory and its overthrow.)

THE INVESTIGATION OF GASES

One of the chief hindrances to progress in chemistry in general and combustion in particular in the seventeenth century was the confusion over gases. This confusion was two-fold: gases could not be collected in a pure state, so could not be clearly distinguished and were generally regarded as various modifications of 'air', and (partly because of these practical difficulties) there was the more general problem of incorporeals (see *Architecture*, ch. 9). Hales, Black and Priestley did much to improve the techniques of collecting, handling and weighing gases and thereby gases were distinguished, identified and recognised as every bit as 'corporeal' as liquids and solids.

Hales

In *Vegetable Staticks* Hales describes 124 experiments on gases and fluids. Many of these are what we now call *destructive distillation,* in which material is heated strongly in the absence of air and the products (gas and liquid) are collected. This work is described in *Architecture,* pp. 232–3, and on p. 234 is the illustration from *Vegetable Staticks* showing the technique Hales used to collect gases free from air. This 'collection over water' was a most important improvement in gas manipulation and was perfected by Priestley. An important criticism of Hales' work was that it was almost exclusively quantitative (Hales was a devoted admirer of Newton[1]). Though this shows a significant shift in emphasis from the too qualitative approach of his predecessors, Hales was perhaps too much concerned with simply 'how much gas' he obtained from a stated weight of material. Had he shown more concern for the *nature* of his products, his significance might have been greater. As it is he is justly extolled for his quantitative approach, for his experimental techniques and for the inspiration he gave to both Black and Priestley.

Black

In 1755 Black published a thesis for a medical doctorate on *Magnesia Alba* (magnesium carbonate). In this he demonstrated simple quantitative relationships in a few well-known chemical reactions. For example, when magnesium carbonate was heated he found a *constant* proportion of the weight was lost. In modern terms: magnesium carbonate = magnesium oxide + carbon dioxide. Furthermore, the 'calcined magnesia' (magnesium oxide) when dissolved in acid yielded the same salt as did the magnesium carbonate, but without effervescence:

magnesium carbonate + acid = magnesium salt + water
+ carbon dioxide
magnesium oxide + acid = magnesium salt + water

Also, if mild alkali (sodium carbonate) was added to the salt produced in the second case above, magnesia alba was precipitated and when dried was found to be very nearly the *same weight* as the sample with which he started. (For further details of Black's experiments see Wightman (pp. 185–7) and Partington (pp. 95–7). The complete works forms no. 1 of the *Alembic Club Reprints*.)

The importance of Black's work lay not so much in the chemistry of the reactions he described (though this was not unimportant at the time) but in his method. Here is the union of quantitative and

[1] Hales' *Vegetable Staticks* bears the imprimatur of Newton as President of the Royal Society. It was given just before Newton's death in 1727.

qualitative and though he describes only a few simple reactions, his use of the balance and his concern over purity of materials have remained a model for chemists. He shows that when magnesia alba is heated, fire does not add the 'principle of causticity' but simply drives off a gas ('fixed air') and this can be detected by the balance. So too can the regeneration of magnesia alba by precipitation with mild alkali. All this shows that chemical reactions, including those in which gases are expelled or absorbed, can be followed by gravimetric methods. The fact that Black in effect described much of the qualitative and quantitative chemistry of carbon dioxide ('fixed air') *without ever collecting it* emphasises this point. (See p. 150 for Black's work on heat.)

Priestley

In a sense, Priestley started where Black left off: his interest in chemistry was first aroused by the collection of the gas (carbon dioxide) given off from the fermenting vats in the brewery next door to his house in Leeds. Details of Priestley's extensive work on gases will be found in many books on eighteenth-century chemistry (see Gibbs' biography; Partington, pp. 110–21; and Wightman, pp. 187–8). It is ironic that Priestley, the staunchest supporter of the phlogiston theory, should have discovered oxygen, by which discovery Lavoisier was to undermine the theory. (There is even greater irony in that he died—a confirmed phlogistonist still—probably of poisoning by carbon monoxide, a gas he sometimes identified with phlogiston!)

It is perhaps of some interest to consider briefly those characteristics that made Priestley one of the most fascinating and controversial eighteenth-century characters (his political and religious controversies might make a good 'fringe' study). He was in all things unconventional. He had a vivid imagination and like most great men was a prodigiously hard worker. He has often been criticised for doing too much, thinking too little and publishing too hastily. He once remarked that had he known any chemistry he would never have made any discoveries (good point for discussion). His discovery of oxygen is his most celebrated 'mistake':

> I cannot, at this distance of time, recollect what it was that I had in view in making this experiment; but I know I had no expectation of the real issue of it. Having acquired a considerable degree of readiness

in making experiments of this kind, a very slight and evanescent motive would be sufficient to induce me to do it. If, however, I had not happened, for some other purpose, to have had a lighted candle before me, I should probably never have made the trial; and the whole train of my future experiments relating to this kind of air might have been prevented.

Priestley's investigations are contained in his *Experiments and Observations on Different Kinds of Air* (1774–91), which runs to six volumes. These are probably the best accounts of the purely empirical method in the history of science.

Priestley is often upheld as the worst sort of scientific reactionary in that he stuck to the phlogiston theory long after Lavoisier had shown it to be false. It is true that he did sometimes draw overhasty and ill-considered conclusions from his experiments, and to stick to phlogiston explanations was certainly contrary to the opinion of the majority. Yet in much of his rearguard action in support of phlogiston, Priestley attacked the oxygen theory of Lavoisier on just grounds and helped to expose several weak points in it. (See p. 151; *Architecture,* pp. 253–5; and White, ch. XII.)

LAVOISIER'S OXYGEN THEORY

There are many admirable accounts and discussions of Lavoisier's experiments and their interpretation by the theory of oxygen. For full treatment see McKie and Guerlac; for a shorter treatment see *Architecture* (pp. 241–56), Partington (ch. VII), Wightman (ch. XVI) and Gillispie (ch. VI).

Perhaps the best brief summary of Lavoisier's significance (and of some of his shortcomings) is that by Partington:

> Lavoisier discovered no new substances, devised no really novel apparatus, and worked out no improved methods of preparation. He was essentially a theorist, and his great merit lay in his capacity of taking over experimental work carried out by others (whose claims he, unfortunately, did not always adequately recognise), and by a rigorous logical procedure, reinforced by his own quantitative experiments, of expounding the true explanation of the results. (p. 122)

In describing Lavoisier's work it is important that several of his experiments be discussed. There is a tendency to regard the 'swan-necked flask experiment' as the single experiment by which phlogiston was overthrown and the oxygen theory established. It was important, of course, but its function was confirmatory for Lavoisier; it was on his earlier simpler experiments, such as the repetition of Boyle's experiment on the calcination of tin (Partington, p. 126), that he first framed his oxygen hypothesis. Due weight must also be given to the influence of his English contemporaries: to Priestley for his criticism in general and for the discovery of oxygen in particular; to Cavendish for his discovery of the composition of water without which the oxygen theory could make little progress, and to Black for a rigorous gravimetric approach to chemistry.

Lavoisier achieved success in his theory partly by his acute insight into the significance of the results of others, but chiefly by the importance he attached to reacting weights. He insisted on the *principle of conservation*. (The law of conservation of mass is sometimes named after him.) In all reactions, losses and gains in weight must balance. This emphasis on *material* changes was at the expense of *energy* changes. His explanations of the latter in terms of 'calorique' are very tentative when compared to the magnificent 'certainty' of his account of material changes (see Wightman, p. 197). Lavoisier's work on chemical nomenclature and classification is discussed at the beginning of the next topic. If desired, it could be included here as a conclusion.

EXPERIMENTS

The principles of combustion and calcination. (Precise details of these experiments will be found in most laboratory manuals.)

(1) Burn a candle in air over water—shows part of the air is consumed. That which remains will not support combustion (phosphorus must be used to get a reduction of one-fifth).

(2) Some materials lose weight on combustion (gaseous oxides), e.g. burn sulphur, carbon or phosphorus in air. Weigh before and after.

(3) Some gain weight (solid oxides), e.g. calcine magnesium. Weigh before and after.

(4) If combustion occurs in *closed vessel* there is no change in weight, e.g. burn phosphorus in closed flask—weigh before and after.

(5) Thermally decompose mercuric oxide, demonstrate oxygen with a splint.

(6) Show ready combustion of materials such as iron filings, candle, charcoal in pure oxygen.

(7) Reduce copper oxide with hydrogen (or coal gas).

II. 6 Classify and divide

The intensive labours of the chemists in the eighteenth century solved many problems, particularly those associated with combustion, calcination and respiration. Lavoisier provided a systematic theory capable of explaining a diversity of phenomena on a minimum of assumptions. Yet although accounts of chemical reactions written in, say, 1780 are more intelligible than much of Boyle's writings, they are still far from 'modern'. The enormous increase in chemical information that had occurred during the eighteenth century had created fresh problems. Chief among these was the need for classification and systematic nomenclature.

LAVOISIER

The importance Lavoisier attached to language in chemistry is illustrated by two extracts taken from his *Elements of Chemistry* (facsimile of first English edition, 1790, published by Dover).

The impossibility of separating the nomenclature of a science from the science itself, is owing to this, that every branch of physical science must consist of three things; the series of facts which are the objects of the science, the ideas which represent these facts, and the words by which these ideas are expressed. Like three impressions of the same seal, the word ought to produce the idea, and the idea to be a picture of the fact. And, as ideas are preserved and communicated by means of words, it necessarily follows that we cannot improve the language of any science without at the same time improving the science itself; neither can we, on the other hand, improve a science, without improving the language or nomenclature which belongs to it. However certain the facts of any science may be, and, however just the ideas we may have formed of these facts, we can only communicate false impressions to others, while we want words by which these may be properly expressed. (Preface, p. xiv)

To those bodies which are formed by the union of several simple substances we gave new names, compounded in such a manner as the nature of the substance directed; but, as the number of double combinations is already very considerable, the only method by which we could avoid confusion, was to divide them into classes. In the natural order of ideas, the name of the class or genus is that which expresses a quality common to a great number of individuals: The name of the species, on the contrary, expresses a quality peculiar to certain individuals only.

These distinctions are not, as some may imagine, merely metaphysical, but are established by Nature. 'A child,' says the Abbé de Condillac, 'is taught to give the name *tree* to the first one which is pointed out to him. The next one he sees presents the same idea, and he gives it the same name. This he does likewise to a third and a fourth, till at last the word *tree,* which he first applied to an individual, comes to be employed by him as the name of a class or a genus, an abstract idea, which comprehends all trees in general. But, when he learns that all trees serve not the same purpose, that they do not all produce the same kind of fruit, he will soon learn to distinguish them by specific and particular names.' This is the logic of all the sciences and is naturally applied to chemistry. (Preface, p. xxvi; see also *Architecture,* p. 243)

These quotations provide the basis for a discussion on the importance of language in science.

Lavoisier, Berthollet, Guyton de Morveau and Fourcroy published their *Methode de Nomenclature Chimique* in 1787 and the ideas therein were incorporated by Lavoisier in his *Traite Elementaire de Chimie* (1788). Table II.2 is the list of elements which appears in the latter work. Lavoisier's definition of an element is given in the *Traite:*

If, by the term *elements,* we mean to express those simple and indivisible atoms of which matter is composed, it is extremely probable we know nothing at all about them; but, if we apply the term *elements,* or *principles of bodies,* to express our idea of the last point which analysis is capable of reaching, we must admit, as elements, all substances into which we are capable, by any means, to reduce bodies by decomposition. (Preface, p. xxiv)

There is not space here to discuss this further but the quotations are sufficient to illustrate Lavoisier's ideas. The preface to the *Traite* is well worth reading in its entirety. It is interesting to note that although Lavoisier's system gained wide and rapid acceptance, the attempts by Lavoisier and his followers to provide a comprehensive matter theory were singularly unfruitful. The justified success of Lavoisier's oxygen theory led the French chemists into an explanatory theory in terms of 'oxyginous, acidifiable, salifiable' and other *principles* which smack very much of the seventeenth century. (This is discussed at length and with examples in Gillispie,

	Noms nouveaux.	*Noms anciens correſpondans.*
Subſtances ſimples qui appartiennent aux trois règnes & qu'on peut regarder comme les élémens des corps.	Lumière........	Lumière.
	Calorique........	Chaleur. Principe de la chaleur. Fluide igné. Feu. Matière du feu & de la chaleur.
	Oxygène.........	Air déphlogiſtiqué. Air empiréal. Air vital. Baſe de l'air vital.
	Azote............	Gaz phlogiſtiqué. Mofete. Baſe de la mofete.
	Hydrogène.......	Gaz inflammable. Baſe du gaz inflammable.
Subſtances ſimples non métalliques oxidables & acidifiables.	Soufre...........	Soufre.
	Phoſphore........	Phoſphore.
	Carbone..........	Charbon pur.
	Radical muriatique.	Inconnu.
	Radical fluorique	Inconnu.
	Radical boracique,.	Inconnu.
Subſtances ſimples métalliques oxidables & acidifiables.	Antimoine........	Antimoine.
	Argent...........	Argent.
	Arſenic...........	Arſenic.
	Biſmuth..........	Biſmuth.
	Cobolt...........	Cobolt.
	Cuivre...........	Cuivre.
	Etain............	Etain.
	Fer.............	Fer.
	Manganèſe.......	Manganèſe.
	Mercure.........	Mercure.
	Molybdène.......	Molybdène.
	Nickel...........	Nickel.
	Or..............	Or.
	Platine..........	Platine.
	Plomb...........	Plomb.
	Tungſtène........	Tungſtene.
	Zinc............	Zinc.
Subſtances ſimples ſalifiables terreuſes.	Chaux...........	Terre calcaire, chaux.
	Magnéſie.........	Magnéſie, baſe du ſel d'Epſom.
	Baryte..........	Barote, terre peſante.
	Alumine.........	Argile, terre de l'alun, baſe de l'alun.
	Silice...........	Terre ſiliceuſe, terre vitrifiable.

TABLE II, 2. LAVOISIER'S LIST OF THE ELEMENTS (1789)

pp. 242–50.) Lavoisier believed in atoms but could not logically apply his beliefs. The successful application of atomism to chemistry came not from the sophisticated French school but from an Englishman, Dalton, who drew far more on Boyle and Newton than on Lavoisier for his inspiration. (This contrast between continental sophistication and English pragmatism appears again on p. 162).

DALTON

[The significance of Dalton in the first half of the nineteenth century has recently been critically re-examined. It seems that the importance traditionally accorded him is exaggerated and that the contributions of Boscovitch, Davy and Faraday have been neglected. For a discussion of this see Knight, *Atoms and Elements*; Brock, *The Atomic Debates*; and Pearce-Williams, *Michael Faraday*.]

The down-to-earth approach to scientific problems alluded to above was characteristic of English scientists in general and of non-conformists in particular. (The influence of non-conformism on English science would make an interesting study as a 'cross-current' involving Newton, Priestley, Cavendish, Dalton and Faraday. Books such as Schofield's *Lunar Society of Birmingham* are helpful.) Dalton was a Quaker and the son of a Lakeland weaver. He was largely self-educated and possessed an ingenious rather than a distinguished mind. He expressed his ideas in a blunt style totally lacking elegance or charm. (It is interesting to compare the sophistication of Lavoisier's *Elements of Chemistry* with Dalton's *New System of Chemical Philosophy*.) At first Dalton had no particular interest in chemistry and he was never very skilful in the laboratory. His chief virtue was an ability to think clearly in models and in his atomic theory he replaced the vague early atomic models with one that was powerful because it was precise.

This atomic theory is the foundation of modern chemistry, yet Dalton came to it by a consideration of purely physical problems associated with gases. Looking back we can appreciate that by this approach he avoided the confusion which beset the French chemists. Having obtained the mental security of a clear physical model he then was able to apply it cautiously to chemical phenomena. Dalton's interest in meteorology led him to wonder at the way in which the gases of the air exist together without affecting each other and, in spite of their different densities, without ever separating into layers. From there he went on to consider the diffusion of gases and the solubility of gases in water.

In his paper 'Theory of the Absorption of Gases by Water' he says:

> The greatest difficulty attending the mechanical hypothesis, arises from different gases observing different laws. Why does water not admit its bulk of every gas alike?—This question I have duly considered, and though I am not yet able to satisfy myself completely, I am nearly persuaded that the circumstance depends upon the weight and number of the ultimate particles of the several gases: those whose particles are lightest and single being least absorbable, and the others more as they increase in weight and complexity. An enquiry into the relative weights of the ultimate particles of bodies is a subject, as far as I know, entirely new: I have lately been prosecuting this enquiry with remarkable success. The principle cannot be entered upon in this paper; but I shall just subjoin the results, as they appear to be ascertained by my experiments. (Quoted in Greenaway, pp. 131–1)

That he 'cannot enter into the principle in this paper' is a pity for he never gave a clear account of precisely how he arrived at his idea—even worse, he later gave several conflicting accounts, none of which tally precisely with his notebooks. The destruction of most of his papers in 1940 has aggravated the problem. (This could lead to a discussion of the problems of historical evidence and the presentation of scientific discoveries.) Dalton's case is particularly interesting for it is clear, in his later accounts, that he was anxious to make the road to his discovery appear more logical than in fact it was. This is common and understandable; the authors of most scientific papers do it, but it gives a misleading impression of the *nature* of scientific discovery. (See Medawar's article in *Experiment*, BBC publications, 1964; also in *The Art of the Soluble*, especially p. 151.)

In his *New System of Chemical Philosophy* (1808) Dalton sets out his ideas at length. It is significant that he does not get on to the interpretation of chemical reactions until page 211. The early part is concerned with the physical problems mentioned above, and in particular with the relationship between heat and matter. It is clear from this part that he owed much of his inspiration to Newton, particularly in the mechanical method of his approach and in his explanation of the mutual intereaction of particles. His concern with caloric illustrates his debt to Lavoisier. With the foundation of atomism firmly laid he then launches into chemical problems—'On Chemical Synthesis'. This occupies only the last ten pages of the first part; the remaining parts were published later and consist of the application of atomic theory to a wide variety of chemical reactions. In the chapter on chemical synthesis Dalton sets out with great precision and clarity (enhanced by his well-known diagrams) the rules governing chemical atomism. He says:

Chemical analysis and synthesis go no further than to the separation of particles one from another, and to their re-union. No new creation or destruction of matter is within the reach of chemical agency. We might as well attempt to introduce a new planet into the solar system, or to annihilate one already in existence; as to create or destroy a particle of hydrogen. All the changes we can produce, consist in separating particles that are in a state of cohesion or combination, and joining those that were previously at a distance.

In all chemical investigations, it has justly been considered an important object to ascertain the relative *weights* of the samples which constitute a compound. But unfortunately the enquiry has terminated here; whereas from the relative weights in the mass, the relative weights of the ultimate particles or atoms of the bodies might have been inferred, from which their number and weight in various other compounds would appear, in order to assist and to guide future investigations, and to correct their results. Now it is one great object of this work, to show the importance and advantage of ascertaining the *relative weights of the ultimate particles, both of simple and compound bodies, the number of simple elementary particles which constitute one compound particle, and the number of less compound particles which enter into the formation of one more compound particle. (New System of Chemical Philosophy,* pp. 212–13)

The rules follow. They are reproduced and discussed in *Architecture* (pp. 260–2) and Greenaway (ch. 7). The important points to emphasise in Dalton's ideas are that his was the first truly scientific atomism; he raised atomism from the speculative status in Newton's 'Queries' to a (for the majority at least) firmly established theory capable of being applied to many and diverse phenomena. He established a sound criterion by which atoms can be distinguished—their relative weights, experimentally determined. Finally, by clearly relating the two fundamental concepts 'atom' and 'element' he gave each concept a far clearer meaning (though not an essentially *new* meaning). From this there followed the idea of atoms of elements combining to form 'compound atoms' or molecules of compounds—again, not a new idea but one which as a result of Dalton's theory became precise and quantitative. As Lavoisier insisted, a clarification of ideas leads to, or rather demands, a more precise expression, and after Dalton chemists were able to express their thoughts, both qualitative and quantitative, in the logical and precise forms of chemical formulae and equations.

Objections to atomism

Dalton made his theory convincing by his discussion of many chemical and physical phenomena in terms of the movement and interaction of atoms. The rapid and fruitful development of matter theory (see below and topic II.7, can be regarded as conclusive evidence of the 'truth' of the theory. (It is a valuable exercise to have the chemists in the class say precisely *why* they believe in atoms. Perhaps the most direct evidence is given by the Brownian motion, see pp. 103 and 109.) But although the majority were convinced, there were sceptics who argued that atomic theory met with such convincing success simply because it was applied only to those problems it *could* solve. (One could discuss here: 'Research is the Art of the Soluble'—see Medawar, *The Art of the Soluble*.) There was little attempt to say *why* there were a limited number of atoms or why two atoms which differed only slightly in relative weights should have radically different chemical properties. There was also the problem which attaches to any atomic theory—that of action at a distance (see p. 106). Akin to this is the problem of the nature of 'chemical bonds'. When Dalton had made atomism so specific and precise such problems as these became more acute.

The misleading persuasiveness of Dalton's theory is well illustrated by the way in which it confirmed the majority in favour of Proust in his celebrated controversy with Berthollet (see *Architecture*, pp. 256–8, for details of this controversy). Philosophical objections were raised by many throughout the nineteenth century. (For a discussion of these see *Architecture*, pp. 262–7 and 298–303; Gillispie, ch. XI; Greenaway, ch. 10; and for Faraday's characteristically shrewd rejection of atomism, see Gillispie, pp. 454–7.) This scepticism is also discussed in topic II.7 (p. 164), and the comment on Lavoisier's oxygen theory (p. 133) might be joined with what is said here in a general discussion of what makes a theory successful, in which could be elaborated the necessities and dangers of selectivity.

Though there was justified scepticism about the relationship between atomism and reality many used Dalton's theory because, whatever its relationship with reality, it was a powerful explanatory hypothesis which clarified much that was known and, more important, directed chemistry and physics along fruitful paths. In the remainder of this topic we shall consider briefly some consequences of the theory in chemistry.

GAY-LUSSAC AND AVOGADRO

One of the most difficult problems in the atomic theory was that of discovering the numbers of atoms involved in chemical reactions. Until such numbers were ascertained the criterion of atomic weight was of very limited use, since different chemists adopted different values. This happened in the first half of the nineteenth century and caused much confusion. Dalton, quite justifiably, insisted that in the absence of evidence to the contrary atoms must be considered as combining in the *simplest* ratio (see his rules, *Architecture,* p. 260). Emphasise that it would have been *unreasonable* of Dalton to assume otherwise. A particularly unfortunate (though logical) consequence of this was that water was given the formula HO and this created great confusion.

Evidence for the number of atoms involved in reactions was afforded by Gay-Lussac's study of the combining volumes. He discovered that in many common reactions gases combine in simple ratios by volume. This suggested to Berzelius and others that these ratios might represent the ratio in which atoms combine, i.e. equal volumes of gases under similar conditions contain equal numbers of reacting particles. This idea led to the well-known paradoxes involving half-atoms, which are faithfully reproduced in many elementary textbooks. It has been said with some justification that Dalton was less than charitable in his attitude to Gay-Lussac's results and their significance, but in fairness it must be pointed out that his objections were not merely a blind rejection of something which apparently contradicted his atomic theory. Quite apart from the 'part atom' paradoxes, he objected to Berzelius' idea on density data (notice the *physical* emphasis again). He found that the density of 'carbonic oxide' (CO) was less than that of oxygen, so if equal volumes contain equal numbers of particles, carbon atom + oxygen atom should be denser than oxygen atom alone. Also Dalton imagined that the spheres of 'caloric' surrounding each atom in a gas were *in contact*, therefore a given number of particles of a compound gas would occupy a greater volume than an equal number of particles of an elemental gas.

The conflict between the evidence of combining weights and that of combining volume was resolved in a very elegant manner by Avogadro in 1811. Details of his reasoning will be found in chemistry text books, but it is better to use his original work in *Alembic Club Reprint, no. 4,* and Cannizzaro's *Revival, no. 18* (see also *Architecture,* p. 272). Avogadro resolved the problem by introducing the idea of a molecule of an element; by suggesting that many of the common elemental gases consist of diatomic molecules (another example of making the simplest assumption—in theory it could be any *even* number of atoms). By this device 'half atoms' and the density objections were removed. It is an interesting and rather sad commentary on chemists of the early nineteenth century that Avogadro's idea was not adopted until revived by Cannizzaro in 1864. Dalton would not accept it for it assumed the truth of Gay-Lussac's evidence. (Dalton's lack of charity towards Gay-Lussac led him to reject not only the interpretation but the *evidence.*) Berzelius could not reconcile it with his successful dualistic theory.

ORGANIC CHEMISTRY

It is often said that by his conversion of ammonium cyanate into urea, Wöhler, in 1828, overthrew the old theory of vitalism and established organic chemistry as a true, albeit now misnamed, branch of chemistry. This is nonsense. It is true that before 1800 it was thought that many substances of animal and vegetable origin could only be made under the influence of a 'vital spirit'. It is true that during the early decades of the nineteenth century chemists questioned the necessity of this 'vital spirit', and that by 1850 organic chemistry as we now understand the term—the chemistry of the compounds of carbon—was firmly established. But Wöhler's conversion was only one step in this slow change. Lavoisier in 1787 had classified all acids, organic and otherwise, together and he clearly thought of organic chemistry as an integral part of chemistry, not as something uniquely related to living organisms. On the other hand, the Liebig–Pasteur controversy in the 1860s (see *Architecture,* pp. 326–30) shows that there were still problems.

This is not the place to discuss further 'vitalism' and the status of organic chemistry. The point is that after Berzelius had shown that organic compounds obey the laws of chemical combination just as do inorganic compounds (though they did not easily fit his 'dualistic' concept), then Dalton's theory could be applied. Its application to organic chemistry, in fact, was one of its great successes. Organic compounds are a very large family of compounds made up of carbon and very few other elements—chiefly hydrogen, oxygen and nitrogen. Many of these compounds are volatile and in consequence their reactions can be investigated by volume measurements as well as by weight. This makes the study of the stoichiometry (i.e. the *numbers* of reacting molecules) of organic reactions rather easier than the majority of inorganic cases and the atomic theory was correspondingly more successful as an explanatory concept. The detailed study of the various ideas of structure in organic chemistry makes difficult reading even for an organic chemist and will not be further pursued here.

One further point of importance arose from the work of Pasteur, Le Bel and Van't Hoff. From their work on optical activity they suggested that the four bonds round the carbon atom are directed towards the corners of a tetrahedron. (Most textbooks of organic chemistry give an account of this.) As a result, organic chemists came to think of molecules in three dimensions in a far more precise way than hitherto, and were concerned far more with the arrangement of atoms in space. The classical work of Kekulé on benzene was important for this reason too (but note Kekulé's caution, *Architecture,* p. 299). Prudent though this caution is, there can be no doubt that the three-dimensional view of organic chemistry and the making of models which it encouraged did much to confirm many in the reality of atoms (see Greenaway, ch. 10 and Partington's *Short History of Chemistry*).

FARADAY

Satisfactory accounts of the work leading to Faraday's laws of electrolysis will be found in books on physical chemistry. The general significance is not always made clear. Faraday discovered that for any *element* the weight liberated by a fixed quantity of electricity was proportional to the weight by which that element undergoes chemical reaction (the equivalent weight). Now equivalent weight is simply related (through valency or 'combining ratio') to the atomic weight. Hence the ability of chemical substances to carry (and hence be deposited by) electricity is an *atomic* effect. The idea of *ions* or charged atoms, which was clarified as a result of Faraday's work, enabled rapid progress to be made in the study of reactions in solution, a part of chemistry which Dalton's rather 'physical' theory never really explained. Maxwell's interpretation of Faraday's results were prophetic (see *Architecture,* p. 278). If desired this work on electrolysis can be expanded into a topic on its own. There is an abundance of experimental work whose interpretation can form a valuable study. Such a topic could be expanded to include the work of Galvani, Volta and Davy on the discovery of current electricity (see Conant, *On Understanding Science,* ch. 3). The definitive biography of Faraday is that by Pearce-Williams. It is particularly important for its reassessment of Faraday's significance in science generally and in matter theory in particular.

Prout

In spite of the fairly widespread scepticism about the physical reality of Dalton's atoms, his theory was a powerful stimulus in chemistry, particularly through the meaning and significance it gave to the accurate determination of combining weights. Some were content to determine these weights even more accurately; these include some great figures of analytical chemistry such as Stas, Dumas, Marignac, Hönigschmidt and Richards. Others looked upon combining weights as a challenge to the imagination and sought new facts through numerical relationships. (It might be interesting to compare generally these two attitudes of mind; they are often found in science.) Prout attempted to show that all atomic weights are whole multiples of hydrogen, hence: 'We may almost consider the protyle of the ancients to be realized in hydrogen.' In other words the hydrogen atom was not merely the arithmetical unit for the chemical scale of weights, but the *physical* unit of which all matter is composed. This 'protyle' attracted the imagination of chemists throughout the nineteenth century, and of course in the 'proton' we have something pretty near to Prout's idea. *At the time* Prout's hypothesis did not withstand the experimental evidence; the analysts showed that many atomic weights were not integers. (See *Architecture,* p. 274, for the censure of Prout.)

MENDELEEF

Several chemists in the first half of the nineteenth century attempted an analysis of atomic weights rather less ambitious than Prout's and, for this reason, rather more fruitful. These chemists concentrated on the relationship between atomic weights and chemical properties. Dobereiner discovered certain groups containing three elements in which the elements were chemically similar and whose atomic weights were in arithmetical progression. De Chancourtois plotted the atomic weights on a helix and found chemically similar elements came under one another. (Details of this will be found in most histories of chemistry: it is instructive to construct a model, see p. 148). Newlands noticed that when the elements are arranged in order of increasing atomic weight then, for the first few at any rate, there is a periodical recurrence in chemical properties: every eighth element is similar. This was his law of octaves for which he was bitterly and unjustly ridiculed. His contemporaries regarded each element as unique and to put elements into families was absurd.

In fact Newlands' method and interpretation were sound, and in 1869 Mendeleef, using essentially the same approach, derived a successful classification of the elements. This is the celebrated 'periodic table', in which similar elements fall in vertical columns and properties change regularly from left to right (see *Architecture,* p. 276). Perhaps the chief reason for Mendeleef's success over Newlands lay in his detailed description of the physical and chemical properties of the elements and the perseverance and care he showed in arranging them. Another difference is that Newlands' system was perhaps too neat and tidy, there seemed to be few difficulties or anomalies. In contrast Mendeleef faced squarely several imperfections and by so doing reaped a rich reward.

The first problem he faced was that of the 'gaps'. In his arrangement he found certain gaps, notably under boron, aluminium and silicon. It seemed unreasonable to suppose that in such a comprehensive and well-ordered system there should be gaps (notice the belief in the 'orderliness' of nature). Mendeleef suggested that these gaps represented the places of hitherto undiscovered elements. (If one remembers the incredulity towards Newlands, Mendeleef's suggestions must for some have bordered on the heretical!) However, this was not merely speculation; Mendeleef went even further by predicting in quite precise terms the properties of the elements. His predictions were based on regular trends he observed in the table. This was in 1871. Within the next fifteen years three elements, scandium, gallium, and germanium were discovered and found to correspond very closely to the properties Mendeleef had predicted. In table II.3 is a comparison of Mendeleef's predictions and the properties of germanium.

Another problem Mendeleef faced was that for a few pairs of elements the order of atomic weights was reversed. Mendeleef was convinced that they must be classified according to their chemical properties and that some explanation would be forthcoming for their 'anomalous' atomic weights. The concept of isotopes vindicated his confidence (see p. 173).

Mendeleef's success in predicting unknown elements convinced many of the essential truth of his idea. At the time it provided an invaluable framework into which chemical facts could be fitted and it suggested new lines of investigation. At the beginning of the twentieth century it formed a comprehensive experimental system which the seemingly rather abstract atomic theories had to explain.

EXPERIMENTS AND DEMONSTRATIONS

Dalton

A simple demonstration of the independent behaviour of gases (strong evidence for an atomic theory) is the diffusion of gases. Pour bromine vapour (not the liquid) down a thistle funnel into a gas jar. At first the bromine will form a dense layer on the bottom,

Table II.3
A comparison of the properties of eka-silicon (predicted in 1871) with those of germanium (discovered in 1885)

Eka-silicon	*Property*	*Germanium*
72	Atomic weight	72·32
5·5	Specific gravity	5·47
13 cm^3	Atomic volume	13·22 cm^3
Dirty grey	Colour	Greyish white
0·073	Specific heat	0·076
White EsO_2	Heating in air	White GeO_2
Slight	Action of acids	None with HCl
EsO_2 with Na	Preparation	GeO_2 with C
K_2EsF_6		K_2GeF_6 with Na
Refractory; sp. gr. 4·7, mol. vol. 22cm^3	Dioxide	Refractory; sp. gr. 4·703, mol. vol. 22·16 cm^3
B.P. 100°C., sp. gr. 1·9, mol. vol. 113 cm^3	Tetrachloride	B.P. 86°C., sp. gr. 1·887, mol. vol. 113·35 cm^3
B.P. 160°C., sp. gr. 0·96	Tetraethyl derivative	B.P. 160°C., sp. gr. 1·00

Taken from Moeller's *Inorganic Chemistry*, p. 117.

but after half an hour it will have diffused throughout the jar. (See also *Nuffield Physics Teachers' Guide,* III, p. 277.)

For the Brownian motion—See p. 109.

Experiments on the laws of chemical combination will be found in elementary textbooks but are rather long and technical for a course of this kind.

Details of demonstrations of Gay-Lussac's law of combining volumes will be found in textbooks.

Model of De Chancourtois' helix

This is useful to emphasise periodicity. Divide the base of a stout cardboard cylinder into sixteen equal parts. From one point at the base draw a line at forty-five degrees. This will describe a helix round the cylinder. If now the atomic weights are marked off on the helix taking each division at the base as a unit of atomic weight, it will be found that chemically similar elements come under one another (this simple rule breaks down beyond scandium).

Organic chemistry

Use molecular models to demonstrate the three-dimensional aspect of chemistry.

II.7 Matter and energy: the classical synthesis

In this topic we trace the development of ideas which led to the 'classical' dichotomy of *matter* (associated with atoms and molecules) and *energy* (associated with electromagnetic waves). The limitations of space here and of time in the classroom, together with the necessarily complex mathematics involved in any detailed treatment, restrict us to a rather general discussion of only the most important ideas. But this we must do, for the ideas involved form the climax of classical physics and also the starting point for twentieth-century developments.

HEAT

By 1800 scientists had a pretty comprehensive experimental knowledge of the manifestations and behaviour of heat. Fahrenheit and Celsius had devised their scales of temperature. Black and Wilcke had clarified the problems associated with the transfer of heat between bodies by the introduction of the ideas of specific and latent heat (see Wolf, II, pp. 178–83. Present-day school experiments on specific and latent heat are essentially those of Black; scientists in the class might explain to others the principles involved.) Watt applied these ideas to the steam engine with singular success. Lavoisier and Laplace had improved calorimetry and studied heat changes attending chemical and biological processes (see Wolf, II, pp. 183–8). Ellicott, Smeaton and others had investigated the expansion of solids.

Several of those working on these problems had remained unconcerned about explanatory theory, but Black employed the currently popular idea of 'matter of heat', an idea which Lavoisier was reluctantly obliged to use and to which he gave the name *caloric*. In his *Elements of Chemistry* (p. 4) he says:

It is difficult to comprehend these phenomena, without admitting them as the effects of a real and material substance, or very subtile fluid, which, insinuating itself between the particles of bodies, separates them from each other; and, even allowing the existence of this fluid to be hypothetical, we shall see in the sequel, that it explains the phenomena of nature in a very satisfactory manner.

Notice the caution with which Lavoisier adopts this idea; he was equally cautious in applying it. This is emphasised because a common and rather unfair criticism of Lavoisier is that he heaped upon this notion of caloric all the features he found so abhorrent in phlogiston. As mentioned in topic II.5 (p. 133), it was largely by ignoring the thermal aspects of the problem of combustion and concentrating on the material changes that Lavoisier's theory was so successful. It was right that Priestley should attack the new theory on its inadequate explanation of thermal changes, but this was a weakness of which Lavoisier, if not all his disciples, was obviously aware.

The chief problem posed by the theory of caloric was that of its apparent lack of weight. By the end of the eighteenth century Newton's ideas were more or less completely accepted, and as a result scientists were becoming very sceptical about imponderable fluids—the overthrow of the phlogiston theory had doubtless encouraged such scepticism. It is true that light, electricity and magnetism were still vaguely regarded as imponderables, but no theory had attributed to these quite the material qualities which the caloric theory bestowed upon heat.

Various efforts had been made during the eighteenth century to detect a difference in weight between a body when hot and when cold—Boerhaave, Buffon, Roebuck and Whitehurst all tried and all failed to detect any difference. The most sensitive and convincing experiments were performed by Benjamin Thompson, Count Rumford, who concluded:

Having determined that water does not acquire or lose weight, upon being changed from a state of *fluidity* to that of *ice*, and vice versa . . . [and] being fully convinced [from the results of the above-mentioned experiments] that if heat be in fact a *substance*, or matter—a fluid *sui generis*, as has been supposed—which, passing from one body to another, and being accumulated, is the immediate cause of phenomena we observe in heated bodies (of which, however, I cannot help entertaining doubts), it must be something so infinitely rare, even in its most condensed state, as to baffle all our attempts to discover its gravity. And, if the opinion, which has been adopted by many of our ablest philosophers, that heat is nothing more than an intestine vibratory motion of the constituent parts of heated bodies, should be well founded, it is clear that the weights of bodies can in no wise be affected by such a motion.

In 1789 Rumford performed his celebrated cannon-boring experiment in which he reamed a cannon barrel with a blunt borer and flamboyantly boiled gallons of water with the heat produced (see Wolf, II, p. 196). This furnished serious problems for the caloric theory. Hitherto heat produced by friction had been accounted for as caloric liberated when material was worn away. In Rumford's experiment only a tiny quantity of filings was produced and an enormous, indeed—and this was the important point—a seemingly unlimited, quantity of heat was produced. These results doubtless confirmed Rumford in his opinions expressed earlier: they did *not* overthrow the caloric theory at once as is commonly reported; like all good ideas, caloric persisted even in the face of adverse experimental evidence for several decades. An important reason for this was that the concept of 'energy' was as yet very vague; the caloric theory quite adequately explained changes of mechanical and heat energy in 1800. When scientists began to consider electrical, magnetic and chemical changes as similar to those involving motion and heat (i.e. as the concept of energy emerged) then the caloric theory became inadequate and was gradually replaced by the more satisfactory molecular–kinetic theory.

Rumford helped revive the alternative explanation of heat, that of motion of constituent particles. In the succeeding decades this was developed into the kinetic theory of matter. (Notice that both the caloric and the kinetic theory involved the idea of motion; in the former it was motion of the particles of the imponderable fluid caloric, in the latter it was the motion of the constituent particles *of the substance*. It was by increasing the emphasis on motion in this second sense that eventually the notion of caloric became redundant.)

In his cannon-boring experiment Rumford also demonstrated a connection between heat and mechanical energy; heat was produced for as long as the borer bored. This led to thermodynamics, which in the hands of Clausius, Boltzmann and Maxwell was united to the kinetic theory. This synthesis formed part of the foundation of classical physics.

THE KINETIC THEORY OF MATTER

As Rumford indicated (p. 151), the idea that heat consists solely in the movement of the constituent particles was not new. It appeared in a speculative form in Democritus' atomism and was revived in the late sixteenth century by, among others, Francis Bacon, who claimed to have deduced it on purely empirical grounds by his method of exhaustive fact collecting (see Wolf, 1, pp. 638–9). Daniel Bernoulli in his *Hydrodynamica* (1735) considered gases to consist of particles in constant motion and that pressure was due to their bombardment of the containing vessel. John Herepath, in 1816, extended this to a dynamical theory of heat. In spite of the quite precise mathematical treatment of Bernoulli these earlier ideas were not generally accepted but they do show a continuous 'kinetic' tradition.

In simple terms the kinetic theory asserts that solids consist of particles (atoms, molecules, ions) vibrating about a fixed position. As the temperature is raised the increase in energy shows as a more vigorous vibration. At a certain temperature, the *melting point* (characteristic for a particular substance), the vibrations become so vigorous that the particles break free from the restricted oscillation and move about, though in a still fairly restricted manner. This is the liquid state and it still defies precise mathematical description. Further increase in temperature leads to more vigorous motion until at the second critical level (the boiling point) the particles break free from their mutual influence and move with rapid and random motion in the space in which they are confined. This qualitative description of the gaseous state is essentially that given by Bernoulli and Herepath and it was in its application to gases that the kinetic theory developed quantitatively.

This was the work of Clausius, who in the 1850s successfully tackled some of the problems that had arisen when the qualitative picture described above was analysed more closely. In order to overcome various difficulties in the application of Newtonian mechanics, the notion of an 'ideal' gas had been introduced: the particles were point masses (occupying no volume), they were separated by very large distances (no mutual attraction or repulsion) and they suffered only perfectly elastic collisions. (One could discuss at this point the value of such 'ideal', unrealisable situations, often created by the scientist to help understand *some aspects* of a problem.) In this ideal picture all the energy of a gas is in the translation (place to place) movements of the particles. Such a picture gives a value for the specific heats of most gases much below those found by experiment.

Clausius showed that if we allow the particles finite size, then a particle—and hence the gas as a whole—can absorb energy by spinning about an internal axis as well as by translational motion, and the specific heat will be higher. Likewise the pressure changes in gases are better explained quantitatively if inter-particle attractions are assumed. Finally, by allowing for appreciable intermolecular collisions, Clausius was able to achieve much better agreement with the experimentally observed rates of diffusion of gases.

In spite of these improvements the theory of Clausius was still unreliable in accounting for the precise behaviour of gases. To overcome this unreliability, Maxwell suggested that Clausius' chief error lay in the assumption that, at a given temperature, the molecules of a particular gas were moving with the same velocity. It seemed to Maxwell that systems containing very large numbers of bodies in random motion might best be treated by the methods of statistics. It is difficult to appreciate quite how startling this suggestion was. We have now accepted not only Maxwell's statistical mechanics but, in this century, a 'probabilistic' interpretation of nature at a deeper level and of a more fundamental kind (see p. 181). But the suggestion that the techniques hitherto applied to such indeterminate matters as games of chance or population trends be applied to the physical world of inanimate atoms must have been, to the nineteenth-century mind, very much like an admission of failure, the cardinal sin. (This might provide an opportunity for discussing the confidence, almost smugness, that marked later Victorian scientists in their attitude to nature. They were confident that nature would be described in a precise, a strictly determinate way. Perhaps historians might comment on a similar attitude shown in politics, art, etc.)

Instead of imagining that the molecules all had the same speed at a given temperature, Maxwell reasoned that in a very large collection of randomly moving molecules, some would be stationary, some moving very fast and the majority would have speeds between these extremes. Boltzmann worked out an expression for the statistical distribution of velocities at a given temperature and obtained graphically the well-known bell-shaped curve. Predictions based on the Maxwell–Boltzmann distribution law agreed very closely with the experiment: in spite of its unpalatability, it worked.

In his paper 'On the Dynamical Theory of Gases' (1866) Maxwell provided a precise description of molecular collision which afforded an explanation of thermodynamics in terms of the statistical mechanics of molecules in motion.

From the kinetic theory of gases were derived all the empirical laws which describe the behaviour of gases. This is a good example of the use of models in science: by the mathematical description of the motion of hypothetical particles, simple empirical laws were linked together in a coherent and single concept—particles in random motion. Also, of course, the empirical laws are experimental evidence in support of the theory. This is a good case through which to discuss the aesthetics of scientific theory: economy—many phenomena explained by one concept; quantification—the whole is rigorously mathematical and the interplay of abstract concept with empirical fact not only enables each to support the other but also suggests new relationships and further fruitful paths of enquiry.

THERMODYNAMICS

In 1800 the application of energy to purely mechanical systems was quite well understood; from a knowledge of effort put in and the 'geometry' of machine employed (levers, pulleys, gears, etc.) the result could be reliably predicted from classical (Newtonian) mechanics. The conservation of energy in such systems was also fairly clear.

In his cannon-boring experiment, Rumford had demonstrated a connection between heat and motion—the energy spent in turning the borer was converted into heat (as we would now say). Observations of this kind, and especially the development of steam power, encouraged scientists to broaden the concept of energy (and its conservation) to such 'non-mechanical' agencies as heat, light, electricity, chemical reaction, and so on.

In 1824 a young French engineer, Sadi Carnot, published a short paper, 'Reflections on the Motive Power of Heat', which was to form the basis of thermodynamics. By considering the cycle of events in an 'ideal' engine, Carnot demonstrated that any engine which could convert all the heat supplied into mechanical work would be a perpetual motion machine, and by operating it in a cycle one could obtain more energy from it than one put into it. (Notice this is a thought experiment—an idealised situation from which valid and useful deductions follow.) In any actual engine only part of the heat supplied is turned into useful work. (We are not concerned here with mechanical inadequacies—a mechanically one hundred per cent efficient engine would still only usefully employ part of the heat supplied.)

If energy was conserved generally, then forms of energy must be converted into each other at fixed rates and it should be possible to determine a constancy in the relationship between two forms of energy. This was achieved by Joule in 1848; he showed that the amount of mechanical work needed to produce a fixed amount of heat was constant. His experiments, which are among the classical researches in physics, will be found described in most physics books on heat.

The problem of generalising these and other studies on energy into a comprehensive conservation law was tackled and largely solved by Mayer and Helmholtz in the 1840s (see Gillispie, pp. 375–93). The conservation of energy became the first law of thermodynamics. But there were still problems, chief among which was one that arose from Carnot's work. If energy is conserved in any process then what changed irreversibly in Carnot's non-ideal engine? It was this irreversible change that necessitated the supply of heat to counter the continual 'running down'.

In his approach to the problem Clausius tried to account for the heat which was always lost. His equations produced a quantity which always increased in any spontaneous process and the increase of this quantity (which was *not* energy) corresponded to the 'non-useful' energy in such processes. He called the quantity *entropy*. Since it is characteristic of all natural (i.e. spontaneous) processes, this increase in entropy denotes a move 'forward' in time in a quite fundamental sense and by means of entropy changes physicists first differentiated time past and time future (Eddington called it 'time's arrow').

A common example of a spontaneous process is the cooling of a body: heat always flows from a hot body to a cold body, never the other way round. 'Heat cannot spontaneously flow from a cold to a hot body' is perhaps the most common statement of the second law of thermodynamics which expresses these ideas of fundamental irreversibility: advance in time and increase in entropy associated with spontaneous processes.

It is interesting to note that in spite of the difficulty which was (and still is) experienced in describing the physical meaning of entropy, Clausius was able to develop thermodynamics successfully by its use. Maxwell's paper of 1866 clarified many thermodynamic notions by expressing them in the 'realistic' terms of molecules in random motion. Perhaps the clearest idea of entropy is obtained by this interpretation. It is thought that the increase in entropy in spontaneous processes corresponds to a decrease in the 'orderliness' of the particles in the system; orderliness is a true imponderable and of course defies conservation laws. The idea that progress (advance in time) is inevitably accompanied by, and is indeed recognised by, an increase in chaos was difficult for the Victorian to swallow; perhaps experiences in this century make it more acceptable!

(The principles of thermodynamics are difficult to grasp and the majority of textbooks will be of only limited value for reference. One of the best popular expositions is Wightman, ch. 23; the books by Ubbelohde in Wightman's bibliography are also useful.)

Light

The problems outlined in the previous sections were concerned with the nature and behaviour of heat (and energy) *within* bodies. Over the same period a series of investigations were carried out on the propagation of energy. Again the final synthesis was achieved by the genius of Maxwell.

Radiant heat had long been recognised as a manifestation of heat. That it behaved in a way similar to light was demonstrated by experiments in which heat and light were reflected by mirrors and brought to a focus by lenses. But because radiant heat was often accompanied by light, the two were not clearly differentiated. In 1774 Buffon focussed the *heat alone* from a hot but not incandescent iron plate and Herschel in 1800 further emphasised the distinction and similarity between heat and light by demonstrating the presence of heat 'beyond the confines of the red' in the spectrum of incandescent bodies (see *Architecture,* p. 281).

Interest in the propagation of what we now call energy arose at the beginning of the nineteenth century out of the revival of the debate on the nature of light. Details of the arguments put forward at different times and by the various protagonists will be found in most physics textbooks. Such accounts, althugh satisfactory for the technical arguments they contain, are often superficial and misleading. A typical outline (omitting details of experiments) might run as follows:

The problem of the nature of light first arose in the seventeenth century. Newton suggested light was a material substance made up of streams of corpuscles. On this theory he was able to account for light travelling in straight lines, reflection and, by means of far-fetched ingenuity (?!), refraction. [Textbooks are anti-Newton in this matter.] In 1678 Huygens suggested light consisted of waves and elegantly explained reflection, refraction and double refraction (but not polarisation[1]). Newton could not accept this theory because light would be expected to bend round objects and it does not. Because of Newton's great reputation everybody believed light to be corpuscular, and Huygens' ideas were neglected for a hundred years. In the 1820's Young, Fresnel and Arago discovered the interference and diffraction of light which cannot be explained on corpuscular theory.

The problem was solved beyond all doubt (and for all time?) when Fizeau and Foucault showed that light travels more slowly in water than in air as the wave theory predicts. The corpuscular theory says it should travel faster in water. Therefore light consists of waves (in ether). Q.E.D.

Before accounts such as this can form any part of a history of ideas several comments need to be made.

The corpuscular theory is as old as the Greeks; Newton adopted and refined it as an explanatory hypothesis. Huygens' theory was indeed mathematically elegant but *on the available evidence* Newton favoured the corpuscular theory and the majority agreed. The reason why so many in the eighteenth century followed Newton's

[1] Polarisation was inexplicable on the wave theory while the waves were considered longitudinal and the ether as a fluid. To overcome this Fresnel suggested that light was a transverse wave motion in an elastic solid ether. (A good example of a change of model leading to a wider and therefore better explanation.)

verdict was not so much his reputation as the power which his genius had given the corpuscular hypothesis as an explanatory hypothesis, not for light alone, but for all phenomena. (Notice economy again—many and various phenomena explained by one concept.) It pervaded all eighteenth-century physical thought; to be 'real' was to be 'material'—made up of particles. This explains why electricity and magnetism were thought of as subtle fluids, and why Lavoisier numbered 'light' and 'caloric' among his elements. The corpuscular theory did 'explain in a very satisfactory manner' (see Lavoisier's quote, p. 151)—rather happier than did the wave theory at the time. It is for this reason that the rather sophisticated ideas of Huygens were ignored.

When we look at the work of Young and Fresnel on interference and diffraction, we see that although their results were substantially the same (for simple accounts see *Architecture*, pp. 283–5 and Wightman, chs. 13 and 14) their approaches were quite different. Young's approach was that which we might expect of the British—'to deliver the master [Newton] from the hands of the epigones of analytical mechanics and *restore* the full Newtonian science of light with its *original sophistication* enhanced by new empirical detail' (Gillispie, p. 407, our italics). Fresnel, who knew no English, had to rely on Marat's translation of *Opticks* and was unable to appreciate Newton's subtlety. (One could discuss here the pitfalls of translation.) His approach was that of a typical continental Cartesian and was the true revival of Huygens.

The experiments of Young and Fresnel and the mathematical elegance of the latter's interpretation did indeed convince the majority that on the evidence *then* available (i.e. 1820) light was best regarded as a wave motion, though a wave motion in or of what remained a problem. The conclusion that light need no longer be regarded as a subtle fluid composed of particles was more acceptable in 1820 because, as we have seen, caloric was already suspect in the minds of some and in the early decades of the nineteenth century there was a gradual acceptance of the fact that to be 'real' need not necessarily mean to be 'material' in the corpuscular sense. It is not really paradoxical that such conclusions became accepted just at the time when the corpuscular theory was achieving its greatest triumphs in both physics and chemistry, for these triumphs were being achieved through the corpuscles being more and more precisely defined in terms of the movement and interactions of atoms and molecules. With this precision came the realisation that light, heat, magnetism and electricity cannot be corpuscular.

ELECTRICITY AND MAGNETISM

These were two further 'subtle fluids' of the eighteenth century. They had been investigated (separately) since the sixteenth century and by 1800 many of their characteristics were known. However, unlike heat and light, the more detailed description of their properties made their nature more obscure. Details will be found of the classical work associated with the names of Coulomb, Ampère and Oersted in textbooks and biographies. Oersted is particularly important for he first demonstrated experimentally the way in which electricity and magnetism were related (see especially Tricker, *Early Electrodynamics*). Conceptually, however, the first really significant work was that of Faraday. He not only investigated exhaustively the relationship between these two phenomena, he also devised a picture or model of electricity and magnetism from which Maxwell was able to derive his electromagnetic wave theory, which together with his kinetic theory of matter formed the foundation of classical physics.

Faraday is justly regarded as one of the greatest experimenters. His most notable achievements were those connected with electricity and his *Experimental Researches in Electricity* is a massive tribute to his industry and ingenuity. We can fairly satisfactorily limit our discussion to three of his discoveries: electromagnetic induction, specific inductive capacity, and the rotation of polarised light in a magnetic field. Precise experimental details are given in most textbooks (or better still, consult *Experimental Researches*), but perhaps a very brief account will be helpful.

Oersted had shown that a wire carrying a current causes the deflection of a magnetic needle. Faraday intuitively felt (see the remarks on Rutherford, p. 175, which apply here also) that there should be a complementary effect whereby a magnetic field in one coil of wire would cause (induce) a current in a second coil. He tried various experimental situations but met with no success even when large currents were employed. Then he observed (another case of chance favouring the prepared mind) that a slight current was induced in the second coil at the moment when the current in the first coil was switched on, and again, but in the opposite direction, when the first circuit was broken. Faraday assumed that the current was only induced as the magnetic field was created and destroyed. This he verified by replacing the first coil by a magnet. While the magnet was still, no current was induced. As the magnet moved, the current appeared. This discovery was of enormous technical importance—the dynamo and electric motor, and hence electrical power generally, depend on it. For our purposes, the significance of electromagnetic induction is that it helped clarify and confirm Faraday's picture of 'magnetic lines of force'. This idea is (and was to Faraday) strongly suggested by the pattern formed when iron filings are sprinkled on a card over a magnet (wires carrying a current also show patterns). Faraday attributed the induced current to the 'cutting' of these lines of force.

The discovery of specific inductive capacity helped Faraday develop these ideas further. He demonstrated that in the well-known phenomenon of a charged body inducing a charge in an initially uncharged body placed near to it, the magnitude of this induced charge depended on the nature of the dielectric, i.e. the substance (usually air) separating them. (He saw in this a strong objection to atomism, see reference on p. 141.) This seemed to indicate that the electricity did not merely jump from the inducer into the induced but somehow *existed already* in the intervening medium.

The third experiment, the rotation of polarised light in a magnetic field, is rather technical to describe (though the scientists in a class should be able to explain the rotation of polarised light by chemical substances). The point is that it demonstrated an interaction and relationship between magnetism and light.

Perhaps the best summary is Faraday's own. Talking of the state or conditions around the magnet he says:

What this state is, or upon what it depends, cannot as yet be declared. It may depend upon the aether, as a ray of light does, and an association has already been shown between light and magnetism. . . . Whether it of necessity requires matter for its sustentation will depend upon what is understood by the term matter. If that is to be confined to ponderable or gravitating substance, then matter is not essential to the physical lines of magnetic force any more than to a ray of light or heat: but if in the assumption of an aether we admit it to be a species of matter, then the lines of force may depend upon some function of it. Experimentally, mere space is magnetic; but then the idea of such mere space must include that of the aether when one is talking on that belief; or if thereafter any other conception of the state or condition of space rise up, it must be admitted into the view of that which just now in relation to experiment is called mere space. On the other hand it is, I think, an established fact, that ponderable matter is not essential to the existence of physical lines of magnetic force. (*Experimental Researches,* III, p. 443)

Faraday's ability to provide such strong physical imagery, such as 'lines of force' and the concept of 'field', and to apply them with such skill arose from his almost total ignorance of mathematics. Although skill in mathematical reasoning has been often reckoned a necessary attribute of a scientist, in this problem and at this time 'while the continental mathematicians were laying down more and more elegant roads in the wrong direction, Faraday spent his time exploring the country to find the right direction' (Wightman, p. 299). This is yet another example of the difference between English and Continental scientists.

Maxwell provided a successful mathematical analysis of Faraday's powerful imagery and the result led to one of the most surprising and successful syntheses of ideas in the history of science. (For a detailed yet non-mathematical treatment of Maxwell's methods see Gillispie, pp. 458–76, and Wightman, pp. 304–16.)

In his memoir of 1856, *On Faraday's Lines of Force,* Maxwell said:

In the following investigation, therefore, the laws established by Faraday will be assumed as true, and it will be shown that by following out his speculations other and more general laws can be deduced from them. If it should then appear that these laws, originally devised to include one set of phenomena, may be generalised so as to extend to phenomena of a different class, these mathematical connections may suggest to physicists the means of establishing physical connections; and thus mere speculation may be turned to account in physical science.

(This is a good quotation for a discussion of the role of models and mathematics in science.)

In order to discover these 'mathematical connections', Maxwell transformed Faraday's non-mathematical point into a seemingly quite different model—a moving fluid. The analogy between the current of electricity in a wire and a fluid flowing in a pipe had been suggested by William Thompson—Lord Kelvin. Now fluid mechanics was a well-established science, and the behaviour of fluids was described by precise mathematical relationships. Maxwell was able to use this mathematical precision to develop further ideas about electromagnetic fields. He subsequently introduced further mechanical features into his model such as cog wheels and even idling cogs (see Wightman, p. 306); by means of these seemingly absurd flights of fancy he was able to obtain deeper and deeper mathematical insight. Finally, the mathematics yielded an equation describing electric and magnetic fields, an equation of the same form as that describing the wave theory of light. This was a most unexpected yet stimulating result, for if two such seemingly disparate phenomena obey the same form of equation then they must be quite closely related. Maxwell was able to calculate an expression for the velocity of his electromagnetic waves and found it agreed closely with the velocity of light determined experimentally by Fizeau. The first 'direct' evidence in support of Maxwell's synthesis (i.e. electricity obeying 'wave' laws) was the detection of oscillatory currents by Hertz in 1887.

Perhaps the most concise expression of the significance of all this is Maxwell's elaboration of the meaning of the title given to his definitive paper on the subject, 'A Dynamical Theory of the Electromagnetic Field':

> The theory I propose may therefore be called a theory of the *Electromagnetic Field,* because it has to do with the space in the neighbourhood of the electric or magnetic bodies, and it may be called a *Dynamical Theory,* because it assumes that in that space there is matter in motion, by which the observed electromagnetic phenomena are produced.

A particularly important point to note here is that in his derivation of this theory, Maxwell employed a new procedure which has been of the greatest use in this century. He sought a mathematical form based only on (Faraday's) facts and not involving *any* presuppositions as to the nature of electricity (whether it be particles, fluid, etc.). He then developed the model purely according to mathematical intuition. As soon as the model or any part of it ceased to be useful it was discarded. The justification was that the *final* result afforded a more elegant (i.e. economical) description of diverse phenomena.

ETHER

Any undulatory theory of propagation seems to pose the problem —undulation in or of what? Maxwell accepted the necessity for some notion of ether (or aether) as had Newton before him but was content to regard its properties as best described by processes occurring in it, and this is what his theory does. He says:

> In speaking of the Energy of the field, however, I wish to be understood literally. All energy is the same as mechanical energy, whether it exists in the form of motion or in that of elasticity, or in any other form. The energy in electromagnetic phenomena is mechanical energy. The only question is, Where does it reside? On the old theories, it resides in the electrified bodies, conducting circuits, and magnets, in the form of an unknown quality called potential energy, or the power of producing certain effects at a distance. On our theory it resides in the electromagnetic field, in the space surrounding the electrified and magnetic bodies, as well as in those bodies themselves, and is in two different forms, which may be described without hypothesis as magnetic polarisation and electric polarisation, or, according to a very probable hypothesis, as the motion and strain of one and the same medium. (Gillispie, p. 476)

For further details of the ether debate and its final fate in the theory of relativity, where it is shown to be *in principle* undetectable, see Gillispie (pp. 490–2 and 507–10).

The kinetic theory of matter gave valuable support to Dalton's ideas and, along with the development in chemistry outlined in the last topic, helped confirm many in the belief in the 'reality' of atoms. Maxwell's electromagnetic theory provided an elegant and rational description of the propagation of energy. Thus the problem of ponderables and imponderables which so beset the eighteenth century was resolved into clearly defined and sharply distinguished particles and waves. But the success of this dichotomy was short-lived. Within a few decades the distinction was once more blurred and the strict classical mind might echo Donne in 'The Anatomy of the World':

> 'Tis all in pieces, all coherence gone.

EXPERIMENTS

Suitable experiments to illustrate much of the subject matter will be found in practical physics textbooks. For a course of this kind and for classes of non-scientists, many of these experiments will be too technical and too time-consuming.

II. 8 Subatomic structure

OBJECTIONS TO DALTON'S ATOMIC THEORY

Dalton's 'billiard ball' atoms and molecules formed a most useful foundation for the study of reactions involving elements and compounds. It was particularly successful as a basis for the rapid development of organic chemistry. The kinetic theory of matter also seemed to provide evidence for the reality of particles of matter. Yet many (including notably Wilhelm Ostwald) remained sceptical. Physicists, especially after the work of Maxwell, were concerned more with space than with matter, and problems of radiation in space seemed to demand a continuum theory. Furthermore, a continuum approach to nature was mathematically more consistent.

In addition to these rather vague and general objections, there occurred throughout the nineteenth century experimental observations which were difficult to account for on the atomic theory. Faraday's work on the decomposition of electrolytes, in which he showed that the carrying of current by a solution was an atomic effect, seemed to indicate that atoms can acquire positive or negative charges in discrete amounts ('atoms of electricity'). This, while supporting the existence of atoms, does not fit very well the 'billiard ball' picture, in which parts of an atom are indistinguishable. The essential correctness of Faraday's observations is demonstrated by the success with which his ideas were developed into electrochemistry by Kohlrausch, Hittorf and others. Their quantitative measurements on the dissociation of ions made even more acute the problems of interpretation on the classical atomic theory.

Again, although Dalton's theory and its important developments by Avogadro and others gave a helpful picture of stoichiometrical chemistry (combining ratios of atoms), and by the concepts of valency and equivalent weight made the quantitative relationships in chemical reactions predictable, it was impossible to say why an element has the valency (or valencies) it has, nor was there any clear picture of what the chemical bond was like. Newton's attractive forces would not do and in any case would not account for specific valencies. (These inadequacies will be all too familiar to those who have tried to teach chemistry on schemes in which such sophisticated ideas as valency and equivalents were introduced long before any satisfactory explanatory theory.) Why there should be a limit to the different types of atoms and precisely why these atoms should conform to the rather intricate pattern demonstrated by Mendeleef were questions which occurred to few nineteenth-century minds; Dalton could have been of no help if they had.

Spectroscopy, the analysis of light emitted by substances, had become a highly quantitative science by about 1870. The spectra emitted by independent atoms (in incandescent vapours) were found to consist of sets of radiation waves (lines in the spectrometer) of quite sharply defined wavelengths (line spectra). (See the very good coloured illustration in Sienko and Plane's *Chemistry*, 2nd ed., McGraw-Hill, p. 50.) The relationships between the large number of lines emitted by an element were discovered by several physicists: Balmer, Lyman, Paschen, Rydberg and others. The equation describing the wavelengths in line spectra suggested that the atom functions as an harmonic oscillator: billiard balls were of no use in explaining this.

These various investigations all indicated that Dalton's concept of the atom was inadequate, yet they did not suggest any better alternative picture. This, together with the enormous areas of physics and chemistry for which the atomic theory was *(and still is)* adequate, perhaps explains why it was so firmly held. (This provides another example for the discussion of the retention of theories in the face of contrary experimental evidence. What follows illustrates why theories are modified: these experiments suggested a picture of the atom of wider applicability.)

CATHODE RAYS

The experimental observations which led to a more satisfactory picture of the atom were those on the electric discharge through gases. As early as 1821 Davy had shown that gases become better conductors when pressure is reduced. The importance of technical improvement is well illustrated by the rapid progress in the study of electric discharge after about 1860, when Heinrich Geissler perfected the technique of sealing metal in glass (necessary for the electrodes in discharge tubes) and also improved the evacuation of the tubes. This is a good example of the benefits of technology in advancing pure science.

The most obvious feature of gaseous discharge was the complex pattern of sparks, but attention was later focussed on the fluorescence produced by the invisible rays coming from the cathode. Details of the experiments whereby the properties of these cathode rays were demonstrated will be found in several of the books in the bibliography; simple apparatus for these demonstrations is readily available. The rays were found to cause fluorescence when they fell on certain substances, to travel in straight lines perpendicular to the cathode, to be cut off by—and hence cast shadows of—metal objects, and to behave as a stream of negatively charged particles—electrons.

The discovery of these characteristics, though interesting, did not indicate any particularly fundamental significance. It was when Sir J. J. Thomson derived the value for the quantity charge/mass of these electrons that it became clear that the electron demanded a rather special explanation. Thomson's method and reasoning behind the determination of e/m will be found in several of the books in the bibliography; it is the result that is significant. He found that $e/m = 1840 \times$ charge/mass of the hydrogen ion (obtained from electrolytic studies). Furthermore, if, as Thomson suspected, the charge on the electron is of similar magnitude to that on the hydrogen ion, then:

$$\text{mass of electron} = \text{mass of hydrogen ion}/1840$$

This is startling, for the mass of the hydrogen ion (which is virtually the same as that of the atom) is on Dalton's theory the smallest mass conceivable, since there is no atom lighter than hydrogen. Thomson's experiments indicate that the electron is not just a bit lighter but of a smaller order of mass altogether. Furthermore, Thomson showed that the properties of electrons, especially e/m values, were independent of the kinds of materials present in electrodes, glass, or gas in the apparatus and hence suggested that these electrons were fundamental and present in all atoms. Thomson was very reluctant to make this revolutionary suggestion and the scientific world was loath to accept it. (See Thomson's words quoted in *Architecture* (p. 308) or the paper from which it is taken:

Philosophical Magazine, October, 1897 (p. 293)—a facsimile reproduction of this and many other important papers of this period will be found in *Classical Scientific Papers,* ed. Wright.)

Since atoms are normally electrically neutral, it seemed reasonable to seek for a corresponding positively charged fundamental particle. Streams of positively charged particles (the positive rays) were detected in discharge tubes but it was then discovered that charge/mass ratios for these particles varied with the gas in the tube. They were therefore not fundamental and were merely the positive residue remaining when electrons were removed from an atom.

Thomson suggested a model of the atom in which the electrons were embedded in an unformed positively charged mass—rather like currants in a bun and it became known as the 'currant bun atom'. Although satisfactory on the available evidence, the 'currant bun' atom could not, in spite of considerable effort by Thomson, explain atomic spectra and it became clear that the 'billiard ball' must be replaced by a radically different model.

For a more satisfactory model of the atom more information was necessary. Rutherford was able to suggest such a model on evidence derived from the use of X-rays and radioactivity which grew out of the earlier investigations into cathode rays.

X-RAYS

In 1895 Wilhelm Röntgen (1845–1903) was investigating the fluorescence of barium platinocyanide in the ultraviolet rays produced when cathode rays strike a metallic surface. He observed that the platinocyanide fluoresced even when it was protected from the ultraviolet radiation by metal foil. The penetrating rays which caused this fluorescence he called X-rays. They attracted much interest and were soon applied in the now familiar technique of X-radiography. In 1912 von Laue, William Bragg and Lawrence Bragg showed that X-rays were electromagnetic radiations of short wavelength by diffracting the rays with the atoms in a crystal. (This was the beginning of X-ray crystallography—a most powerful means of investigating structures.)

As will be seen later, Moseley, under Rutherford's guidance, was to clarify a most important feature of atomic structure by the analysis of X-ray wavelengths.

RADIOACTIVITY

In 1896 Henri Becquerel (1852–1908) wondered if Röntgen's X-rays were produced when chemical substances fluoresce in sunlight—a phenomenon he had been investigating. He wrapped crystals of various fluorescent chemicals in protective foil and exposed them to sunlight. On developing the plates he found that those used with uranium showed fogging, indicating that penetrating rays were produced. Then by chance he developed some plates which had been near uranium minerals but these samples of uranium had not been exposed to sunlight. Here was a new phenomenon—penetrating rays produced by uranium without any external stimulus. The subsequent elucidation of the nature of this radioactivity, including the heroic efforts of Marie Curie, will not be elaborated here (see bibliography). For our purposes it is enough to say that radioactivity was found chiefly in the elements heavier than lead, and that as a result of the early work of Curie and Rutherford the radiation was found to consist of α-particles, which had a double positive charge and an atomic weight of 4; β-particles, which were fast electrons; and γ-rays, which were very short wavelength electromagnetic radiations.

α-particle scattering

Rutherford observed that when α-particles moved through air some were sharply deflected. He was able to observe this by means of the 'cloud chamber' devised by C. T. R. Wilson in 1912. This device is a splendid example of the application of relatively simple principles to produce a tool of enormous value in the study of often complicated phenomena (see Wilson's papers, *Phil. Trans. Roy. Soc. London,* **A.188,** 265–307 (1897); *Proc. Roy. Soc.,* **A.87,** 277–92 (1912). These are reprinted in *Classical Papers,* pp. 311–74.)

In 1906 Rutherford had observed that α-particles penetrate thin metal foil but in so doing spread out slightly, and instead of a sharp spot on a photographic plate behind the foil, he obtained a diffuse area. These observations suggested that the α-particles were deflected when they encountered matter and it occurred to Rutherford that a more careful study of α-particle scattering might give some clue as to how matter was arranged inside the atom. The experiments were performed chiefly by his students, Nuttall,

Geiger and Marsden (see *Classical Papers*, pp. 143–82) and produced surprising results. Most α-particles passed through a sheet of metal atoms without any deflection or were deflected through only a very small angle. A small but definite proportion, however, were deflected through a very large angle and left the collision chamber very near to the spot where they entered. The fact that the majority were undeflected seemed to indicate that the atom was largely empty space (this destroyed the 'billiard ball' and the 'currant bun'), but even more surprising was the fate of the few deflected through a large angle. Rutherford expressed this surprise in characteristically robust terms: 'It was about as credible as if you had fired a 15-inch shell at a piece of tissue paper and it came back and hit you.' In 1914, as a result of these observations, Rutherford put forward his now celebrated model of the atom (see *Classical Papers*, pp. 206–45). At the centre was a small, dense, positively charged nucleus; small because so few α-particles encountered it, dense because virtually all the mass of the atom was centred there, and positively charged because the deflections of the α-particles indicated repulsion (+repels+) rather than collision. The nucleus is surrounded by electrons, equal in number to the positive charge on the nucleus, and moving in orbits round the nucleus like planets moving round the sun, the electrostatic attraction being analogous to gravitational attraction.

This model, though it has been considerably 'refined', remained the model on which all subsequent developments were based (but more recent 'refinements' seem to deny the possibility—or, indeed, the necessity—for a model in this sense; see p. 189.)

It is convenient now to consider separately subsequent developments of the nucleus and the electrons.

THE NUCLEUS

From their experiments Geiger and Marsden were unable to give precisely the magnitude of the positive charge of the nucleus of a given atom, but it seemed to be about half the atomic weight. In 1914 Moseley, yet another of Rutherford's illustrious students, analysed the wavelengths of X-rays emitted by electron bombardment of a variety of metals He found that these wavelengths fell into a series, corresponding to the sequence of elements when arranged in order of increasing atomic weights. The position of an element in this sequence was defined by its atomic number. 'We have here proof that there is in the atom a fundamental quantity which increases by regular steps as we pass from one element to the next. This quantity can only be the charge on the central positive nucleus, of the existence of which we already have definite proof' (Moseley, *Phil. Mag.*, **(6) 27**, 1031 (1913) and in *Classical Papers*, p. 189). In spite of all this activity there was no indication of a counterpart of the electron, a positive fundamental particle. The lightest positive particle known was the nucleus of hydrogen (atomic number 1), and in 1914 Rutherford suggested that this be regarded as 'the long-sought positive electron'. Though it is nearly two thousand times heavier than the electron, subsequent investigations showed that, as in most of his guesses, Rutherford was right. This fundamental particle was called the proton after the 'protyle' of Prout who, in 1815 and on *no substantial evidence*, had suggested that all atoms were made of units corresponding to the hydrogen atom (see p. 145). The discovery of the positron, the true positive electron, in 1932, in no way invalidated Rutherford's suggestion, for the positron is not a fundamental particle in the atom (see p. 186).

In 1920, in his Bakerian lecture to the Royal Society, Rutherford entertained another 'inspired guess':

> Under some conditions . . . it may be possible for an electron to combine much more closely with the hydrogen nucleus (a proton), forming a kind of neutral doublet. Such an atom would have novel properties. Its external (electric) field would be practically zero . . . and consequently it should be able to move freely through matter. Its presence would probably be difficult to detect.

This hypothetical particle was called the neutron. In 1924 (Sir) James Chadwick wrote to Rutherford: 'I think we shall have to make a real search for the neutron. I believe I have a scheme which may just work. . . .' Eight years later the scheme did work: Chadwick detected the neutron when beryllium was bombarded with high energy protons.

The main features of the nucleus were now clear. It consists of protons (number equals atomic number) and neutrons such that the sum of these makes up the atomic weight. The nucleus of a particular element is characterised by the number of protons. The number of neutrons can vary within certain narrow limits; thus a sample of a particular element may consist of atoms with nuclei of slightly different weight. Such atoms may be regarded as forms of an element; because they have the same number of protons in their nuclei, they will have the same number of electrons around the nucleus. It is the electrons which govern chemical characteristics since only these are involved in chemical reactions. Hence these forms of an element will be *chemically* indistinguishable and will occupy the *same position* in Mendeleef's periodic table. For this reason they are called isotopes (Gk. *iso* = same, *topos* = place).

The existence of isotopes, demonstrated by Aston and Soddy, explained many problems, especially fractional atomic weights and also the anomalous behaviour of certain pairs of elements in Mendeleef's classification (see p. 146). Atomic weights measured by classical means give the *average* weight of the isotopes—this depends on the weights of the isotopes and their relative abundance. (See textbooks for amplification of this. It might be interesting to speculate how Mendeleef would have fared if nature had played the game rather differently and created average isotopic weights out of sequence with atomic number—the really important characteristic—for many more elements.)

THE ELECTRONS—THE BOHR ATOM

Since it is the electrons which govern chemical properties, Rutherford's picture of the atom was not really much use to the chemist, for although it gives the number of electrons and accounts for their existence in the atom, it gives no clue as to their arrangement. Niels Bohr, while working under Rutherford at Manchester, tackled this problem of electron arrangement. He derived his theory of electron orbits from a consideration of line spectra (p. 167) and applied the quantum theory (p. 177).

Rutherford's picture of orbiting electrons was unsatisfactory since, according to classical electrodynamics, such electrons should emit radiation continuously. Now in so doing they would lose energy and hence would spiral into the nucleus; Rutherford's atoms would therefore be unstable and have only a transitory existence. Clearly this is not the case; atoms are stable and when they do emit radiation they do so in the sharp clearly defined wavelengths which make up the line spectra. As an answer to this, Bohr suggested that the electrons rotate about the nucleus in orbits of fixed energy (stationary states). When in such an orbit the electron, in spite of classical electrodynamics, does not emit or absorb energy, so the atom will be stable. Changes in energy are achieved by the electron jumping either to an orbit of higher energy further away from the nucleus (absorption of energy), or to one of lower energy nearer the nucleus (emission of energy). Since the electron cannot exist between orbits, energy will be emitted in discrete packets (quanta) corresponding to jumps to adjacent orbits or to more remote orbits. As is explained in topic II.9 (p. 178), the energy of a quantum is proportional to its frequency (and inversely to its wavelength), hence the discrete wavelengths in line spectra.

Space does not permit a full discussion of the subsequent developments of the Bohr atom, chiefly by Sommerfeld and Pauli, but as a result of their work we know the precise number of electrons in the various orbits and on this more refined model was explained the arrangement of elements in Mendeleef's table. Thus the Bohr atom provides a theoretical basis for Mendeleef's empirical table and, vice versa, the periodic table stands as powerful experimental evidence for the Bohr atom. The Bohr model also gives a satisfactory picture of the chemical bond. From a knowledge of the electron configuration of very stable atoms (the rare gases) Lewis, Langmuir and Kossel were able to derive the electronic theory of valency by means of which, at least in a general way, they were able to account for chemical reactions. Again details will be found in all intermediate chemistry textbooks. In a class composed of scientists and non-scientists it might be a good exercise to have the former explain to the latter these relatively simple yet fundamentally important applications of the Bohr atom.

In spite of its immediate and wide success in chemistry, the Bohr model soon met with serious objections. These led to the changes outlined in topic II.9; and relatively simple and mentally 'tangible' models gave way to 'pictures' of ever-deepening mathematical abstraction.

RUTHERFORD

Perhaps as an endpiece to this discussion of the foundation of atomic structure a few comments on Rutherford will be appropriate. He was, perhaps more than any other, the 'guiding light' of this pioneer work. Many of the now great names in this period were his students at Manchester between 1907 and 1919—Geiger, Marsden, Darwin, Bohr, Moseley, Andrade, Chadwick, Fajans, Hevesy and Nuttall. He was a great experimenter and a great director of research. He mistrusted 'abstract' theory and was very dubious about some of Bohr's ideas and even more dubious about the 'mathematical abstractions' discussed in topic II.9. He once said of theorists: '. . . they play games with their symbols, but we turn out the real solid facts of nature.' Perhaps he is noted chiefly for his intuition in devising helpful experiments and particularly in making shrewd, farseeing and often correct guesses. We have noted some of these in connection with the proton and neutron. Some may say this sort of intuition is God's gift to the few, but perhaps a more rational reason is to be found in the fact that, like Newton and Faraday, Rutherford was a prodigiously hard worker and, if asked how he made his discoveries, he would probably have echoed Newton's answer: 'By always thinking unto them.'

(See Andrade, *Rutherford,* especially pp. 208–13, and *Rutherford at Manchester* which also contains reprints of several of his papers.)

EXPERIMENTS AND DEMONSTRATIONS

It is doubtful if there will be time or facilities for doing much in connection with this topic. Sets of discharge tubes are available for showing the properties of cathode rays. With counting equipment and a few sealed radioactive sources, the properties of radioactive emission and the random nature of the process can be demonstrated; several good laboratory manuals are available. (Teachers unfamiliar with the handling of radioactive materials should seek the advice and assistance of someone qualified.)

II. 9 The quantum theory

In topics II.6 and II.7 were discussed the two chief ideas in nineteenth-century physical science: the atomism of matter and the wave nature of energy. The work of Dalton, Mendeleef, Clausius and Maxwell reflects the great firmness with which Newtonian or classical mechanics were held and, in turn, their triumphs represent the surest evidence of the reliability of Newton's ideas when applied to a world of experience.

In the last topic we saw how the idea of the indivisibility of atoms and the immutability of elements was undermined and replaced by a more complex picture, which explained the rather large number of elements in terms of a few fundamental particles. The most satisfactory model of the atom from the point of view of the chemist was that of Bohr, in which the positively charged nucleus is surrounded by electrons which are constrained to occupy and move in clearly defined constant energy or stationary states. The arrangement of electrons in these orbits accounted in a general way for Mendeleef's periodic table and the pattern in atomic spectra, and also a reasonably clear and acceptable picture of the nature of interatomic bonding.

It was mentioned there that in deriving this picture of fixed energy orbits Bohr was attempting to combine classical mechanics and quantum theory. Though the result provided chemists with a useful working model, it was unacceptable to many physicists, and the subsequent criticism (no one was more critical than Bohr himself) led to fruitful development of quantum theory and also to the disappearance of classical mechanics from the atomic world.

THE BREAKDOWN OF CLASSICAL RADIATION THEORY

In order to understand a little more about this most important contemporary idea of the quantum, we must go back to the closing years of the nineteenth century. In the 1870s Lummer and Pringsheim had demonstrated experimentally the distribution of energy among the wavelengths of a radiation body. The result when shown graphically was a characteristic 'hump' curve shown below. As the temperature increases the maximum rises and moves into the shorter wavelengths, i.e. more of the energy is concentrated in the shorter wavelengths. The sequence of events when an electric fire is switched on illustrates this: at low temperatures the peak is in the infra-red, we feel the heat but see no colour in the element; as the temperature rises the peak moves into the visible and the element glows red; further increase in temperature causes the wire to become white hot as more and more energy is concentrated in the shorter (blue) wavelengths. All attempts to explain this distribution of energy in terms of classical theory failed;

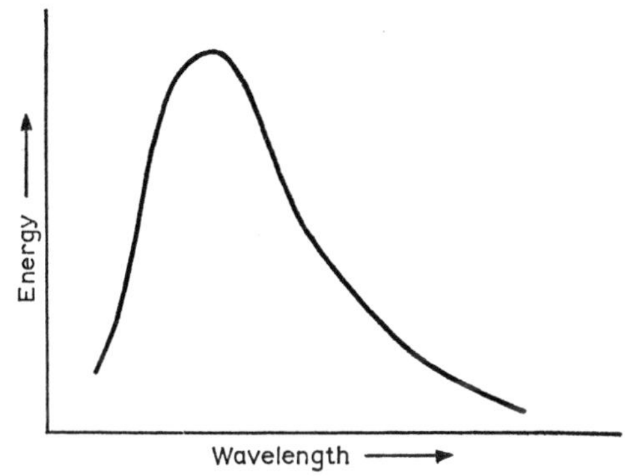

the application by Jeans of the hitherto successful ideas of Maxwell and Boltzmann led to the absurd conclusion that whatever the temperature all the energy would be concentrated in the shortest wavelength. This is the 'ultraviolet catastrophe'. Experimental results show that this is not the case—indeed our daily experience shows it, for as Gamow points out, if the classical prediction were true then the kitchen oven would deliver a lethal dose of short wave radiation when the door was opened!

THE IDEA OF QUANTA

In a paper delivered in the last month of the last century Planck put forward an entirely new idea of energy which accounted satisfactorily for the energy distribution in radiation. The novelty of Planck's idea was that energy was not emitted as a continuous stream (this was a fundamental assumption of classical physics) but in discrete energy packets. Each packet had a well-defined energy proportional to the frequency of the electromagnetic wave:

$$\text{energy} = \text{constant} \times \text{frequency}$$

The packets are called *quanta* and the constant is a universal constant called Planck's constant and is given the symbol h.

When applied to the distribution of the energy from, say, an electric fire at about 800°C, this theory predicts that only very few quanta of high energy (high frequency, short wavelength) will be present; the bulk of the energy will be emitted as the lower frequency quanta of the red region of the spectrum. Though the energy per quantum is less than those of higher frequency, there are more 'red' quanta emitted, hence most energy of the fire is in the red region.

When working on the problem of radiation, Planck realised that it did not involve the nature of the emitting body, only the relationship between energy and temperature. A satisfactory solution to the problem, then, 'promised a deeper insight into the relation between energy and temperature, which is the principal problem of thermodynamics and therefore also of the whole field of molecular physics' (Nobel address, 'On the Origin and Development of Quantum Theory', Clarendon Press, Oxford, 1922).

Evidence in support of the quantum theory

In his Nobel address Planck mentions the difficulty he had in giving physical meaning to his equations; this discreteness of energy was the only reasonable interpretation. Planck himself was reluctant to suggest such a revolutionary idea and others were vociferous in rejecting it. Clearly, it was desirable that some experimental evidence other than radiation energy distribution should be forthcoming. In 1905 Einstein used the quantum theory with great success to explain the photoelectric effect. When electromagnetic radiation strikes certain metals electrons are ejected. Now, on classical wave theory, the energy of a wave is expressed by its amplitude and not by its frequency; the greater the amplitude, the greater the energy (intensity). Sound affords a useful illustration: the greater the amplitude the louder the note; changing frequency (and wavelength) merely changes the pitch. We would expect, therefore, that radiation of sufficient amplitude would provide sufficient energy to dislodge the electrons and that the effect would be independent of the frequency. The opposite is found to be the case: for a given metal only radiation above a certain frequency (the threshold frequency) will give the effect at all. When the frequency is increased above this, the electrons are ejected with greater energy; increasing the amplitude (intensity) of the radiation increases the number of electrons ejected. Radiation of frequency lower than the threshold frequency will not move electrons no matter how great the intensity.

Planck had suggested that radiation was emitted in discrete quanta; Einstein extended this and to account for the photoelectric effect suggested that radiation was absorbed as quanta. If the energy is given by Planck's equation, then it is proportional to frequency and only quanta (or photons, as Einstein called them) of energy equal to that required to kick out an electron will produce the photoelectric effect. If the energy is greater (higher frequency) then the electron will be kicked out more violently; if the high radiation is of greater intensity there will be more photons and more electrons will be ejected. If the energy of the quantum of radiation is less than that corresponding to the threshold frequency, no electrons will be affected, no matter how many photons impinge.

This was valuable evidence in support of Planck's idea, and since radiation had been shown to be both emitted and absorbed as quanta it was reasonable to suppose that energy existed as quanta all the time—an assumption Planck was initially very loath to make. Planck's reticence during this exciting period is worth emphasising. Quantum theory was his idea alone and it was a revolutionary idea in that it involved a complete break with classical physics, yet Planck was reluctant to make the break complete until experimental evidence demanded it.

Experimental justification for quantum theory followed. Einstein successfully accounted for the variation of specific heat with temperature which, like the photoelectric effect, defied explanation in terms of classical physics (see *Revolution in Physics*, p. 119). Further support for the quantum theory was afforded by the *Compton effect*. When X-rays strike an electron the frequency of the scattered X-rays is less than that of the incident beam. Now, the lower frequency (longer wavelength) means that the energy of the scattered quanta is less than that before collision. Compton showed that this drop in energy is equal to the energy gained by the electron in the collision. This cannot be explained by classical theory. Finally there was Bohr's theory of the quantisation of electron energy levels in the atom (see p. 173).

THE WAVE-PARTICLE PROBLEM: DE BROGLIE

It would seem from all this that the 'particle' view of matter and energy (for quanta are in some ways like 'atoms' of energy) was becoming firmly established at the expense of waves. But all problems were by no means solved with the application of the theory; as always, many new problems were raised. Perhaps the most important were those raised by Bohr's theory of energy levels. If energy is emitted by an electron jumping from a higher to a lower energy level, how are we to imagine the jump occurring? Rutherford, who remained very critical of quantum theory, pointed out that it seemed as if the electron must know beforehand where it is going to stop (see *Architecture,* p. 320). Also Bohr's theory demands that an electron *may not exist* between two quantum levels, but how are we to picture the transition? These and the many other more technical objections to Bohr's theory arose because it was a *compromise* between classical ideas (particles describing precisely defined orbits) and quantum ideas (permitted and forbidden energy zones).

It was this unhappy compromise which encouraged Louis de Broglie in his development of wave mechanics, from which all classical ideas are absent:

> The essentially discontinuous nature of quantisation expressed by the appearance of whole numbers, the quantum numbers, in the formulae presented a strange contrast with the continuous nature of the motions visualised by the old dynamics, either Newtonian or Einsteinian. From all the evidence, we had to succeed in setting up the new mechanics where quantum ideas would come to take their place at the very base of the doctrine and not be added on *ad hoc* as in the old quantum theory.

In chapter 8 of *The Revolution of Physics,* from which the above quotation is taken (pp. 175–6), de Broglie tells the story of his invention of wave mechanics. His account is not mathematical and is eminently readable. Briefly, de Broglie suggested that just as waves show some particle properties, so particles might show wave characteristics (a desire for symmetry played a part in this inspiration). He derived his expression for the wavelength associated with a particle from the fundamental equations of Einstein and Planck. (The derivation is simple and will be found in any book on wave mechanics.) Schrödinger applied wave mechanics to Bohr's atom and explained the stationary states as sets of standing waves: electrons can only exist in those regions around the nucleus which can accommodate a whole number of waves (see diagram in Gamow's *Biography of Physics,* p. 251). On this picture electron movements within the atom consist in a change in the wave patterns corresponding to the loss or gain of energy. (For useful musical analogies see *Architecture,* p. 327, and *Quanta and Reality,* pp. 29–30; *Architecture,* pp. 325–8, gives a good account of this idea.) Thus the problem of electron jumps between energy levels was solved and the paradox and arbitrariness of the original Bohr atom were removed. De Broglie solved most of the problems associated with the Bohr compromise. The theory was intellectually more satisfying since it was based entirely on the new wave mechanics.

As with Planck's quantum theory many were sceptical of de Boglie's ideas, for that waves should be associated with electrons (and in principle—*any* particle) was a strange idea. Fortunately experimental verification was soon forthcoming. Davisson and Germer, in 1926, and G. P. Thomson, in 1927, obtained diffraction patterns (characteristic of waves) where a beam of electrons impinged on a metal surface (see *Revolution in Physics,* pp. 174–8).

SCHRÖDINGER

Schrödinger's wave equation (for a derivation and description of this see Linnett (ch. 1), though there is little to be gained from a rigorous treatment in this subject) represents an attempt to apply de Broglie's ideas about an electron to systems of electrons about a nucleus. The equation is a differential equation in terms of what Schrödinger called a wave function Ψ. Ψ is formally similar to the amplitude of the wave, but in this context it is difficult to give it physical meaning. By applying the limitations imposed on an electron in the atomic environment only certain solutions of the equation in terms of Ψ are possible. These are called the 'proper values' (eigenvalues). Now although Ψ cannot be given any readily comprehensible physical meaning, Ψ^2 is a measure of the probability of finding an electron. Although probability is not very easy to visualise, if we abandon the electron as a particle—as a fixed point in space—and imagine it 'smeared' over a volume of space, then the probability (Ψ^2) corresponds to the density of the smear. (For sound, amplitude implies loudness or *intensity*.) If we plot, in two dimensions, the distribution of Ψ^2 about a nucleus we get a 'picture' of the electron in relation to the nucleus. Perhaps an example will help. Let us consider the simplest atom, hydrogen. This consists of a nucleus of one proton surrounded by an electron

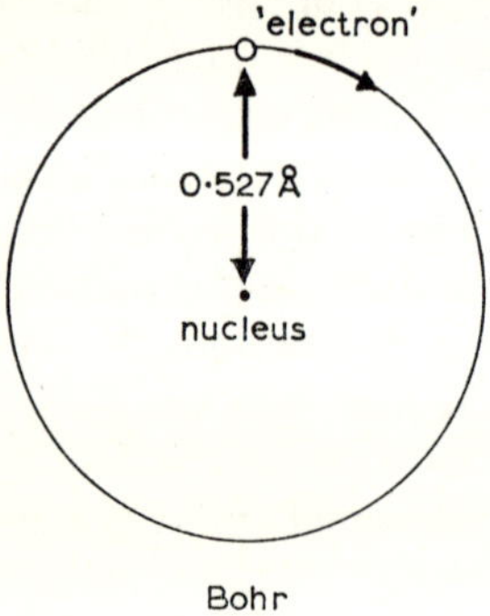

Fig.II.1

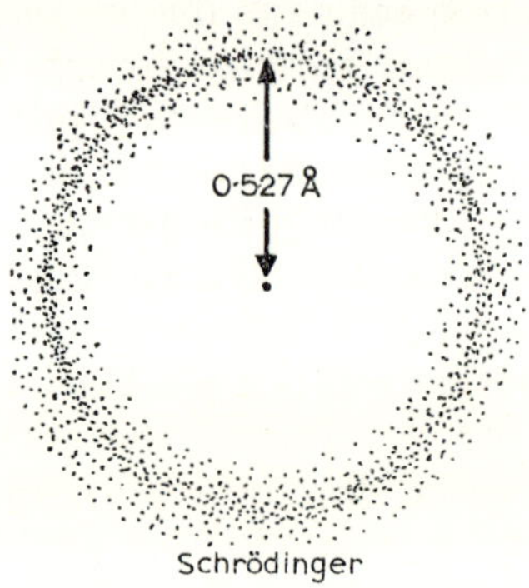

Fig.II.2

in a given position. The Bohr picture is shown in figure II.1. Bohr was able to calculate the radius of the orbit and obtained reasonable agreement with experiment. Now the result of plotting Ψ^2 (actually $4\pi r^2\Psi^2$) about the nucleus is shown in figure II.2. Notice that the classical planetary electron has been abandoned and is replaced by a 'cloud' round the nucleus. The density of the cloud increases as we move away from the nucleus, then decreases again. (N.B.: this density is not the density of a large number of corpuscular electrons but that of a single electron 'smeared out' according to Schrödinger's equations.) Now, the most dense area (in reality this is a three-dimensional volume) is seen to be at a distance from the nucleus marked x. This means that though we may find an electron very near the nucleus or at a very large distance from it ('thin cloud'), *the most probable area (volume)* in which an electron will be found is that at about the distance x from the nucleus. By calculation it is found for the hydrogen atom that this corresponds very closely with the Bohr radius ($0{\cdot}529 \times 10^{-8}$ cm) and with the radius of the hydrogen atom found experimentally.

So Schrödinger's ideas seem to be verified, and moreover at the time gave a far more satisfactory picture of the atom for both the chemist and the physicist. By 'satisfactory' we mean an account of the atom which not only explained adequately all that was known but suggested further lines of investigation which have proved to be fruitful. This increase in our ability to *understand* electrons was achieved at the cost of making it more difficult to *visualise* the inside of the atom.

HEISENBERG

At about the same time that Schrödinger was devising wave mechanics as a means of describing the electron, Heisenberg, a student under Bohr in Copenhagen, derived a theory (quantum mechanics) which gave results very similar to those of Schrödinger's wave mechanics but which was based on entirely different assumptions. (Schrödinger later showed that although the starting point and the method of these two techniques were so different, there was considerable mathematical equivalence between them and hence the similar results were not just a coincidence.)

It is very difficult to give a 'popular' comprehensible account of Heisenberg's method. He was impressed by the discontinuity inherent in the old (already old!) quantum theory and in his derivation he resolved physical quantities such as position, velocity, force, etc., into sets of numbers called matrices. This was a very bold step for it effectively removed all physical meaning from the qualities but it was justified, Heisenberg argued, since there was no point in discussing electrons in atoms in terms of position, velocity, etc., when in such systems these are not observables. This was why Planck's theory and Bohr's application of it to atoms were unsatisfactory. The 'observables' in the atomic system are the radiations resulting from electron transitions between stationary states and Heisenberg introduced into his calculations only such observables as these. (This is called a 'phenomenological' approach, in that it attempts to eliminate from physical theory all that does not correspond to observable phenomena.) Once the information was in matrices then the powerful, if complicated, methods of matrix algebra could be applied. (This might be an opportunity for the mathematicians to explain some of the features of matrix algebra, e.g. non-commutative multiplication—when $a \times b \neq b \times a$, etc.) With these conditions and by these methods, Heisenberg's theory satisfactorily explains all known quantum phenomena.

The uncertainty principle

(See Gamow's *Biography of Physics*, p. 254.)

This was one of the most important consequences of Heisenberg's approach to matter theory. It arises out of his refusal to construct his theory on large-scale observables such as position and velocity which do not apply to the atomic scale. It states that in principle we cannot simultaneously exactly measure the position and velocity (or related properties) of an electron of similar small body.

The usual expression for the uncertainty is

$$\triangle d \times \triangle v \doteqdot \frac{h}{m}$$

where $\triangle d$ is uncertainty in position, $\triangle v$ is uncertainty in velocity, h is Planck's constant (10^{-27} c.g.s. units), and m is mass of body (10^{-27} g.) for the electron.

Since h is constant and m finite then $\triangle d$ and $\triangle v$ cannot both equal 0 at the same time. If we try to devise an experiment whereby the position is accurately determined ($\triangle d \rightarrow 0$) then the uncertainty in velocity ($\triangle v$) becomes very large and vice versa, hence complete knowledge, in the classical sense, is impossible. Notice that for a heavy body h/m decreases and for ordinary bodies the uncertainties can be neglected. Perhaps an example will help. If a body weighs 1 gram then $h/m \doteqdot 10^{-27}$. If we measure its position to within one millionth of a centimetre (10^{-6}) then $\triangle d = 10^{-6}$ and

$$10^{-6} \times \triangle v \doteqdot 10^{-27}$$

$$\triangle v \doteqdot \frac{10^{-27}}{10^{-6}} = 10^{-21}$$

Hence Heisenberg allows us an error in the velocity of 1 cm in 10^{21} seconds, or about 1 cm in 10^{15} years, or about one millionth of a centimetre per thousand years! Q.E.D.

But on the atomic and subatomic scale with which Heisenberg was concerned, masses are very small, hence h/m is large (about 1 for the electron) and the uncertainties are large.

Notice that this is not an expression of the limitation of experiment, nor does it that perhaps one day we shall be able to measure these things exactly. It states that such measurements are impossible in theory. This imprecision or uncertainty in defining both the position and state of an electron is in a sense a property of the electron and is independent of the skill and methods devoted to investigation, for however great the skill and however refined the methods, the act of measurement will always interfere with the thing which is being measured.

Heisenberg illustrated this by a 'thought experiment', wherein he imagines the attempt to locate the position of an electron and to measure its velocity with a 'high powered microscope' (see *Biography of Physics,* p. 255). These thought experiments in which imaginary—but, in terms of classical physics, justified—situations are created in order to demonstrate a theory or refute an objection, are popular in twentieth-century discussions—(see also the Einstein–Bohr controversy in *Biography of Physics,* p. 260).

The uncertainty in the description of an electron prevents the accurate prediction of its position at any future time no matter how accurately its past history is known. The most we can do is express its future behaviour in terms of probabilities; herein lies the agreement between quantum and wave mechanics. The term 'probability' applied to subatomic phenomena by Schrödinger and Heisenberg is fundamentally different from the normal use of the term. Perhaps the most successful application of the normal idca of probability was in the description of the kinetic behaviour of molecules in the nineteenth century (see p. 154). In this case the random behaviour of an almost inconceivable number of particles was expressed in manageable terms which enabled valid predictions to be made. For such a system the behaviour of each molecule was describable in classical terms and its future could be predicted; to do this for a very large collection of bodies, though *in principle* possible, is in practice impossible and the statistical mechanics of Maxwell and Boltzmann afforded an adequate and reliable description of such a system. The point to emphasise is that in this case the probability applied to the consequences of a very large number of events; each individual event was determinate. Now the 'probabilities' in quantum and wave mechanics are a convenient way of expressing the inherent uncertainty of any *individual* event in subatomic systems. Such systems are therefore fundamentally *in*determinate. This was philosophically one of the most startling and, for some (e.g. Einstein), alarming things about the new mechanics. Any lingering hope that some day, somehow, quantum mechanics will become nice and determinate in the classical sense was dispelled when Neuman proved that as classical and quantum mechanics are now understood, any return to determinacy is not only improbable but is, on our *present* ideas, impossible—the probability laws of the new mechanics are incompatible with a hidden determinism.

THE POSITRON

A further demonstration of the usefulness of quantum mechanics was provided by Dirac in 1929. He successfully united relativistic and quantum mechanics. (He incorporated the former in order to cover the case of fast electrons where relativistic considerations are important.) Perhaps the most startling feature of his calculations was a set of solutions corresponding to a world of negative energy and negative mass (the dream of the phlogistonists in topic II.5!). This world lies almost without the imagination; it is a world in which in order to move objects forward we will pull them back! However fantastic such a world may seem, Dirac showed that fundamental particles from such a world could have a definite, if transitory, existence in our world.

The 'electrons' of this weird 'negative' world would be identical to the electrons of our 'positive' world except that they would have a positive charge. Thus Dirac predicted the existence of the long-sought-for positive electron, the *positron* (originally the positron was conceived as the hole left in the negative world—of which we are blissfully unaware—when an ordinary negatively charged electron is removed). Because our world abounds in electrons, the positron very soon meets an electron (the electron sinks into the hole again!), mutual annihilation occurs and the energy is released as short wave radiation. (See *Biography of Physics,* pp. 216–19.)

All this was highly speculative and improbable until, in 1932, Anderson observed tracks in a cloud chamber (see p. 170) which corresponded exactly to a positron.[1] In 1955, Segre and Chamberlain detected negatively charged protons and thus arose the idea of anti-matter—a strange mirror world of our own made up of anti-electrons (positrons), anti-protons and anti-neutrons (discovered in 1956). Whenever an anti-particle meets its counterpart, a real particle, annihilation occurs with the release of energy. It is now believed that supernovae—such as the sudden bright star which so impressed Tycho—may be the radiation produced from the annihilation of colliding positive and negative galaxies.

OMEGA-MINUS

Perhaps one of the most spectacular recent examples of the application of mathematical formalism in the investigation of nature was the prediction and discovery of the omega-minus baryon (a subnuclear particle). It was found by Salam and others that when certain characteristic properties of such particles were plotted, the points formed simple regular geometrical patterns. Of particular interest was the decad formed by families of ten closely related particles (Plato and Pythagoras would rejoice at this!). In an article in the *New Scientist* of 20 February 1964, Matthews explains the significance of the particles and of these geometrical patterns to which they conform. He shows that a particular family of nine fall into a decad pattern with a gap at one vertex (figure II.3). It seemed reasonable to suppose that this gap represented an undiscovered particle (omega-minus) whose properties could be predicted. (This recalls Mendeleef's prediction of elements, p. 148.)

In the *New Scientist* a week later (27 February 1964) the discovery of this particle was reported by Strutt using the proton synchrotron at Brookhaven (see bubble-chamber photograph and interpretation, figure II.4).

These articles are well worth having and provide useful material for discussion. A further point for discussion could be the trouble which ensued when some objected that such an important discovery should first appear in a 'popular' journal, long before it appeared in the more erudite (and less read) journals!

[1] It is interesting to note that after Anderson's interpretation of positron tracks, several other workers reported that they had observed such tracks previously but had attributed them to experimental errors of some kind—chance favours the prepared mind!

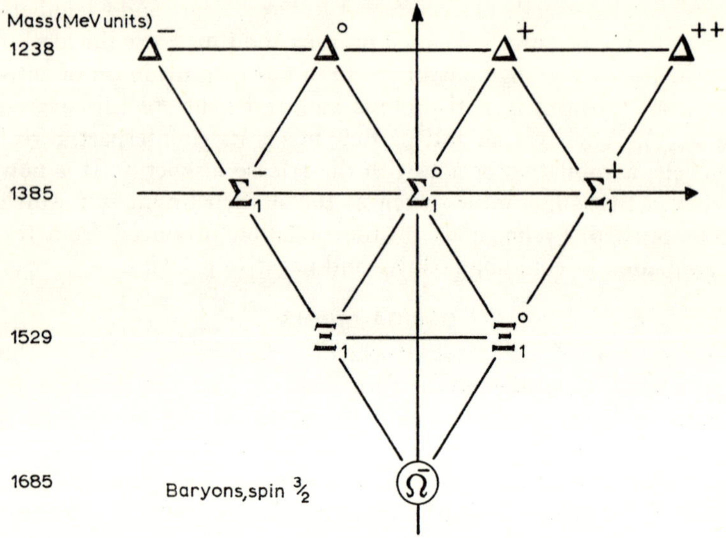

Fig. II. 3

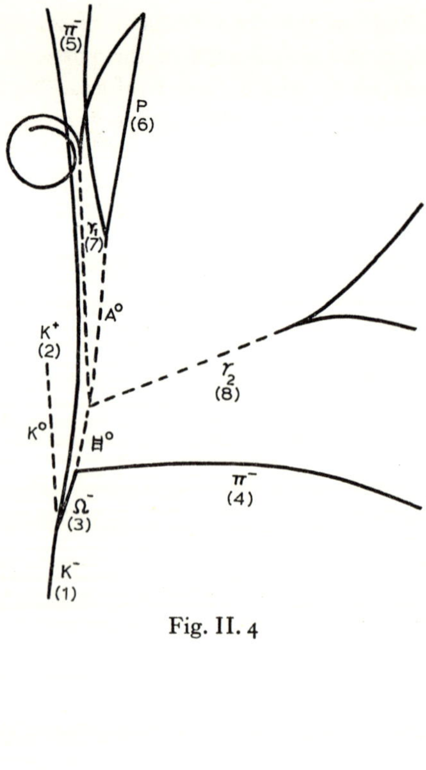

Fig. II. 4

QUO TENDIMUS?

The characteristic feature of the matter theories of this century is that they do not always arise out of past experience and problems but are freely invented and often move away from physical realities, yet ultimately produce more adequate and experimentally verifiable results. This is achieved at the cost of our being even less able to get a clear mental picture or model of reality. To those unprepared or unable to visualise reality solely in terms of the solutions of mathematical equations, this is a serious disability, but one which we must accept.

Perhaps more alarming to anyone who has followed the development of matter theory from the Greeks to the present day is the belief, firmly held by many, that quantum mechanics represents, albeit at the moment imperfectly, the *final* framework in which our views on matter will be expressed. The uncertainty and indeterminacy, however unpalatable to the classically nurtured mind, are of the essence of reality, and any effort to get over this by digging yet deeper into a subquantum world is not only unjustified on the present evidence but is in principle nonsensical. Bohm and others, however, insist that a subquantum dig is worthwhile. *Quanta and Reality* amplifies much that has been mentioned in this topic and also provides much stimulating thought on the wisdom or otherwise of attempts to enter the subquantum world (see also Hesse, ch. 10). Dirac, in *Principles of Quantum Mechanics* (Oxford, 1958), concludes:

It would seem that we have followed as far as possible the path of logical development of the ideas of quantum mechanics as they are at present understood. The difficulties, being of a profound character, can be removed only by some drastic change in the foundations of the theory, probably a change as drastic as the passage of Bohr's orbit theory to the present quantum mechanics.

III Scientific ideas in Biology

DAVID HUGHES-EVANS

Lack of space has presented a problem over the selection of topics. A selection has been made, and no doubt some may disagree with it, while others may regret that topics such as Greek biology have not been included. True, the work of Aristotle, or the work of Galen, could have made excellent topics in themselves. In the section that follows, however, the major part belongs to the postmicroscopic era. Perhaps some explanation or justification for the selection is necessary.

In the first place, the aim here has been to trace the development of ideas in certain areas, rather than concern with one particular period. Though many of the ideas originated with the Greeks, since that time they have undergone development and change. It is true, too, that only in the last three or four hundred years have hypotheses been rigorously tested, and one must bear in mind that the growth of knowledge follows an exponential curve.

In the second place, the inclusion of the history of ideas as part of a programme of liberal education demands that we keep in mind certain educational and social objectives. These are concerned with the development of citizenship, and the maintenance of scientific literacy. In addition to showing how science works, it is essential to discuss the place of the scientist as a man in this pattern. Also one must take the students to the frontiers of knowledge, the point where scientific prediction gives way to scientific speculation.

(It has not always been convenient to define technical terms in the text: a glossary is provided on page 244.)

III. 1 William Harvey

In his book, *Science, its Method and its Philosophy,* G. Bruniston Brown states that the meaning of science has changed from that of 'knowledge', to mean 'the method of obtaining knowledge'. Perhaps this very change emphasises the evolution that has taken place in science itself. The meaning of 'science' is important in any discussion involving the history of science. The meaning we give to it can decide to a very considerable degree our views of 'Greek science' or 'pre-science', and the extent to which many early ideas may be regarded as 'scientific ideas'. In considering the eclipse in learning during the Dark Ages, one can perhaps regard modern science as having been influenced by, rather than being a direct continuation of, Greek science.

Changes in both the physical and the biological sciences were gradual, for the authoritative acceptance of the writings of Aristotle and Galen lingered on for a very long time, and scientific matters were still debated without reference to experiment. The reaction of scientists of the sixteenth and early seventeenth centuries was against the blind adherence that had developed towards these writings, and not against the writings themselves.

Early changes in biology can be seen in the work of Leonard Fuchs (1501–66), *De Historia Stirpium* (1542); and of Andreas Vesalius (1514–64), *De Humani Corporis Fabrica* (1543). The first was in the field of botany, the second of human anatomy. The early changes could perhaps be expected in the descriptive sciences of botany and anatomy, for these would be most closely bound up with art. One can see in the books mentioned above a major difference from earlier works: the information is not provided from ancient texts, but from observations or from dissections. This tendency has developed gradually from the time of Leonardo da Vinci (1442–1519), and can be clearly seen in the wood-cuts of Otto Brunfels of Mainz (1489–1534) and of Jerome Bock of Heiderbach (1498–1554). These works, however, differed from modern works in their presentation. Vesalius is always, as it were, thinking of the body as a whole, and the work is not sectionalised. Neither do we find any of the comparative approach so characteristic of modern biology.

This is an interesting subject, and one that can be developed for class discussion. Reference can be made to the chapter 'The Role of Art in the Renaissance' in *Critical Problems in the History of Science,* edited by Clagett; or to the chapter 'Rebirth of Inquiry' in Singer's *A History of Biology.*

Nearly a century later, and having a more profound impact, appeared the work of William Harvey (1578–1657), *De Motu Cordis* (1628). The change of attitude is well illustrated in Harvey's own words: 'I profess to learn and teach anatomy, not from books but from dissections, not from the tenets of Philosophers but from the fabric of nature.'

It was this change in the attitude of scientists which took place in Europe at this time that resulted in the formation of scientific societies, such as the Royal Society (1660) in this country and the Academie des Sciences (1666) in France. To quote the charter (1662) of the Royal Society, it was intended for 'Further promoting by the authority of experiments the sciences of natural things and of useful arts.' (See McKie, ch. 1, *The Royal Society—its Origins and Founders.*)

There is no better place to begin a study of the history of biology than with Harvey's work on the circulation of the blood. Harvey's work must be regarded as one of the first instances of classical experimental biology. To commence at this point is not to minimise the tremendous debt of modern biology to the Greeks, but to highlight the importance of the change which took place at this time; this change in scientific thought and attitude may be regarded as the origin of modern biology.

THE CIRCULATION OF THE BLOOD

When considering the work of Harvey, it is essential to make reference to conditions prevailing at the time. The works of Galen (A.D. 130–200) and others were held to be almost infallible and the activities of scientists were restricted almost entirely to the study of these works. There was still a widespread respect for ancient writings and any person departing from their general acceptance, as Harvey did, did so at risk to his reputation.

What is Harvey's place in the history of biology? Many of his terms had already been mentioned by others, and much of the evidence Harvey used was already known. On reading Harvey's work, one is struck by his constant references to older writers, in particular to Aristotle and Galen, and the debt he owes them. He also reviews the physiological views of the times. (Singer deals with this aspect in his *A Short History of Anatomy and Physiology from the Greeks to Harvey,* pp. 175–80.) Hippocrates (*c.* 460–*c.* 370 B.C.) considered the pulse as being due to the movements of blood vessels, and Herophilus (*fl.* 280 B.C.) and Erasistratus (*fl.* 280 B.C.) had distinguished veins and arteries. Galen had described the valves in the heart, although his explanation of their action was unsatisfactory. Vesalius had observed the distribution of blood vessels in the lungs, while Michael Servetus (1511–53) and Realdo Columbo (1516–59) had given a clear account of the minor (pulmonary) circulation. Fabricius of Aquapendente (1537–1619) had described and demonstrated the valves in the veins. It can also be argued that Andrea Cesalpino (1519–1603) mentioned the fact that blood circulated rather than oscillated in the ebb and flow manner described by Galen. In 1876, a statue was erected in Rome to Cesalpino, with the inscription that he was the first to discover the circulation of the blood. However, was his meaning of circulation the meaning that Harvey gave to circulation, or was it blood leaving the heart during the day, returning to the heart at night? It is very important in assessing the work of any scientist to read the work as a whole and not to be tempted to give modern interpretation to an isolated passage.

The criteria used in assessing the historical importance of any scientific work should be discussed, and the question may be asked: 'Is it right to credit Harvey with the discovery of the circulation of blood?' What is the evidence? In the first place Harvey added to the knowledge available on the heart and circulation. In the second place he put forward a well-founded theory on the circulation of the blood. Lastly, it was as the result of his work that the theory became generally accepted.

GALEN

Galen is difficult to read. His ideas on the blood system need to be correlated with his acceptance of the four elements, and with the functions he attributes to the lungs, and the notion of pneuma. A short clear account of the ideas of Galen are given in *Science since 1500* by H. T. Pledge:

Galen's doctrine ran roughly as follows. Blood formed and charged with 'natural spirits' in the liver, ebbs and flows under the heart's action between the liver and the right ventricle of the heart. But a little penetrates the (valveless) septum into the left ventricle, there to meet air from the lungs (in breathing we draw the basic principle of life, 'pneuma', from the general or world spirit) and become 'vital spirits', which are distributed by the arteries—little blood and much spirit. Some goes to the brain, where it becomes 'animal spirits', distributed by the nerves (then thought hollow to permit this), and so to muscles and elsewhere. (p. 27)

In Galen's picture the veins carried the nutritive spirits and the arteries carried the vital spirits. Distribution was brought about by an ebb and flow of the blood. The diagrammatic representation of Galen's physiological system, found in Singer's *A Short History of Anatomy and Physiology from the Greeks to Harvey* (p. 61), will help the student to understand the ideas of Galen. It may also be possible to show at this stage the film *William Harvey and the Circulation of Blood*. At the beginning of this film there are excellent animated diagrams to illustrate Galen's thoughts on the 'Motion of blood'. The film could also be used as an introduction to the work of Harvey. The traditional view of Galen's physiological system represented here has been the subject of much research. It is suggested that there was ebb and flow on the arterial but not on the venous side. (See the paper by D. Fleming in *Isis*, **46**, 14 (1955), where it is also shown how the tradition view arose and was perpetuated.)

HARVEY'S WORK

The reading of Harvey's work is different. It is a good idea to read one of the translations, and see how Harvey presents his data and reasoning, and to see how the whole idea of circulation is presented long before the term is mentioned. There is no difficulty in reading Harvey's work for the language is that of modern science.

In the first thirteen chapters of his book, Harvey discusses the evidence for the circulation of blood, evidence such as that provided by the structure of the heart and the application of ligatures. Then, in chapter 14, he briefly and clearly sums up. The following quotation is from the Everyman's Library edition of Harvey's *The Circulation of the Blood*:

Conclusion of the demonstration of the circulation.

And now I may be allowed to give in brief my view of the circulation of the blood, and to propose it for general adoption.

Since all things, both argument and ocular demonstration show that blood passes through the lungs and heart by the action of the (auricles and) ventricles, and is sent for distribution to all parts of the body, where it makes its way into the veins and pores of the flesh, and then flows by veins from the circumference on every side to the centre from the lesser to the greater veins, and is then finally discharged into the vena cava and the right auricle of the heart, and this in such a quantity or such afflux and reflux, thither by the arteries, hither by the veins, as cannot possibly be supplied by the ingesta, and is much greater than can be required for mere purposes of nutrition; it is absolutely necessary to conclude that the blood in the animal body is impelled in a circle, and is in a state of ceaseless motion; that this is the act or function which the heart performs by means of its pulse; and that it is the sole and only end of the motion and contraction of the heart. (p. 85)

Having considered Harvey's presentation it is equally necessary to consider the observations he made. He carried out dissections of live animals, as others had done before. He dissected many different species, and was more fortunate than his predecessors in that he dissected cold-blooded animals. The heartbeat of cold-blooded animals is sufficiently slow to enable a mechanical analysis to be made. Harvey also asked the right kind of questions:

What does the heart do in animals that do not have lungs?

What happens in the foetus before it can use its lungs?

The acceptance of Harvey's work had far-reaching effects. How much understanding of general physiology would have been possible without an understanding of circulation? There are also many important aspects of medicine which need an understanding of circulation.

Harvey's discovery seems simple to us today. Harvey, however, was unable to provide the final verification for his theory. He entertained various ideas as to the link between veins and arteries but came to no firm conclusions. The existence of capillaries was demonstrated by Marcello Malpighi (1628–94). This was Malpighi's first major contribution with the use of the microscope.

In 1660 he saw the blood flowing from artery to vein through the capillaries.

The development of Harvey's work had to await the perfection of the lens, so that these capillaries could be seen. With Harvey's work the limit of understanding possible without the microscope had been reached.

THE TWO PHILOSOPHIES

The growth of physiology more than any other factor brought to an end the influence of Aristotelian biology, and with it the decline in the philosophy of vitalism, a doctrine that life cannot be fully explained in terms of purely physical factors. To Aristotle, the extra factor was the 'psyche'; in fact he posited three kinds of psyche—'vegetable', 'animal' and 'rational'. Animals possessed animal psyche in addition to vegetable psyche, and this gave them the characteristics of movement and responsiveness which plants do not possess. Man differed from other animals in having a rational psyche. (Taylor discusses this subject in his book *Aristotle*, pp. 75–7.)

At the time of Kepler, Galileo and Harvey, we see the growth of mechanistic philosophy. No one did more than Réné Descartes (1596–1650) to establish this doctrine. To Descartes, it was possible to explain life by the same laws as were used to describe non-living matter. However, by introducing his rational soul, he did not apply his philosophy in toto to man.

From the time of Harvey, physiologists have in general adopted a mechanistic approach. This can be seen in the mechanical analysis of Giovanni Alphonso Borelli (1608–79), in the chemical interpretations of Sylvius (1614–72), and at a more elementary level in the plant physiology of Jan Baptiste van Helmont (1577–1644). Not all scientists, however, adopted this philosophy. Georg Ernst Stahl (1660–1734) returned to the philosophy of vitalism, his 'sensitive soul' being not very different from the psyche suggested by Aristotle. Another notable adherent of vitalism was the embryologist, Hans Driesch (1867–1941), who referred to a vital principle in the development of the embryo. Nonetheless it is true to say that in biology there has been a trend towards the mechanistic approach, and that this tradition has continued up to the present time. (It is interesting that the purely mechanistic approach to physical science is now under attack; see topics II.2. For further information, see 'Essentials of Vital Activity' and II.9 in Singer's *A History of Biology;* also Goodfield's *Growth of Scientific Psysiology.*

One of the major contributions of Harvey's physiology was in replacing vital explanations by explanations more akin to those in the physical sciences. The acceptance of a mechanistic philosophy has, however, made one assumption obligatory. If we are to explain biology in terms of physics and chemistry, then advances in the physical sciences must precede biological development. Modern advances in the fields of cellular and molecular biology seem to support this view.

Is the principle of vitalism alien to scientific thinking? A vitalist would postulate 'vital force' when a problem appears without solution. A scientist cannot truly accept this, he must behave as though all problems can be solved by the scientific method. Even if there is a finite limit to our knowledge, a scientist's attitude must be: How can we possibly know when such a state has been reached? Mechanistic philosophy, on the other hand, states that there can be no such thing as a vital force; by definition it cannot exist. A scientist, therefore, is somewhere in between: although his approach is mechanistic it is not necessary for him to accept the whole doctrine of the mechanists. (For a further discussion of the mechanist–vitalist dispute see Goodfield, *The Growth of Scientific Physiology.*)

III. 2 The origin of life

THE DOCTRINE OF SPONTANEOUS GENERATION

To find the origin of this doctrine, one has to go back at least as far as Anaximander in the sixth century B.C. He taught that animals had their origin in the slime of the sea. Aristotle (384–322 B.C.) also believed in this concept. Suitable breeding material could give rise, according to Aristotle, to worms, insects, fish, frogs and mice. In Aristotle's *Historia Animalium,* there are passages such as this:

Some insects are derived from insects of the same kind. Other insects are not derived from living parents but they are formed spontaneously. Some of the latter are formed from dew falling on leaves; others in timber, either green or dry; some in the hair of animals; and still others in the flesh or in the excrements.

These beliefs, although they appear to us odd, were based on cautious observations. When one considers the degree to which Aristotle was recognised as an authority, it is easy to understand how the belief in this concept became so widespread. This point is well brought out by Keosian in his book *The Origin of Life*:

The scholars of the Middle Ages, steeped in superstitions and with only the badly translated and half-understood writings of Aristotle to guide them, added some fantastic embellishments to the established fact of spontaneous generation. Sometimes, as in the case of the goose tree, convenience was served better than science, for if geese grew on trees they were proper food for meatless fast days. Travellers returned from the Orient with tales about a vegetable lamb found in gourdlike fruit.

Alchemists, though failing to turn base metal into precious ones, were convinced that even a homunculus, a miniature manikin, could be produced in a flask. The famous Paracelsus, as late as the sixteenth century, gave a detailed prescription for this astonishing feat, and also reported on his observations of various instances where mice, frogs, eels, etc., were generated spontaneously from water, air, and decaying wood. Is it any wonder then that Shakespeare, who wrote about timeless truths, also repeated the old wives' tales and views of his time by letting his characters in *Anthony and Cleopatra*, *Julius Caesar*, and *Hamlet* tell about the generation of animals from moist mud, decaying hides and hair, rotting meat, etc., activated by the heat of the sun? (pp. 9–10). [On p. 10 there is a diagram representing a goose tree, and on p. 11, a diagram of a vegetable lamb.]

In the seventeenth century these ideas were challenged. One of the first to do this was Francesco Redi (1626–98). Although his approach was scientific, he was nevertheless unable to convince those who firmly believed in spontaneous generation. Redi's experiments are well worth considering. By covering the mouth of a jar containing pieces of meat and eel, while leaving another flask uncovered, he was able to observe flies constantly entering and leaving the uncovered jar. It was the meat in this jar that became wormy. Redi was able to state: 'Thus the flesh of dead animals cannot engender worms unless the eggs of the living be deposited therein.' Perhaps his inability to convince others is not to be wondered at, for in spite of his observations and deductions he himself still believed that gall insects were spontaneously generated!

During this time also we have the microscopic work of Antony van Leeuwenhoek (1632–1723). He considered that his microscopic discoveries 'must surely convince all of the absurdity of these old opinions that living creatures can be produced from corruption or putrefaction' (see chapter 'The Development of Microscopic Observation' in Hughes).

The work of Redi and Leeuwenhoek had opened the door, and for the next hundred years biologists debated this question. Was it biogenesis or was it abiogenesis? The debate had two highlights: in the middle of the eighteenth century the Needham–Spallanzani debate, and then the Pouchet–Pasteur debate in the mid-nineteenth century.

Chapter 2, 'Life from Life', in *Biological Science: An Inquiry into Life* (B.S.C.S.), provides a clear and interesting introduction to this section. The account given of the experiments of Redi, Joblot and Pasteur provide useful student reading material. On page 5 of the student laboratory Guide to this book is an exercise, 'Early Experiments in Spontaneous Generation'; this provides a very suitable laboratory demonstration.

Needham versus Spallanzani

John Turberville Needham (1713–81) was an abiogenesist, while Lazzaro Spallanzani (1729–99) was a biogenesist. The debate started by Needham repeating experiments similar to those carried out earlier by Louis Joblot (1645–1723). Joblot had worked with hay infusion and in his experiments had used a closed and an open vessel. By including the second closed vessel he must have suspected that the air was a source of contamination. Needham used mutton gravy; his results were the reverse of Joblot's and supported the concept of spontaneous generation.

Spallanzani repeated Needham's experiments but was unable to confirm his results. Spallanzani used eight containers, all of which he boiled. Four of these he closed with cork, the remaining four had airtight seals. Growth was observed in the first four containers, but not so in the second four.

The argument was on: the abiogenesists maintained that for spontaneous generation to take place, air was necessary, while the biogenesists held that air was the source of contamination. (See Singer's *A History of Biology*, pp. 440–3.)

Pouchet versus Pasteur

F. A. Pouchet (1800–72) was an abiogenesist, while Louis Pasteur (1822–95) was a biogenesist. Pouchet knew of the objections of the abiogenesists, and to overcome these objections he used artificial air in his experiments. This should not be contaminated as natural air would be. The work was challenged by Pasteur. Both were in fact carrying out similar experiments. Pasteur, however, was working with infusions of yeast, while Pouchet was working with hay infusions. In the debate that followed the French Academy of Science played an important role. The result of Pasteur's work finally resulted in the downfall of the doctrine of spontaneous generation. An account of Pasteur's work can be found in *Harvard Case Histories in Experimental Science* (cases 6 and 7), *Milestones in Microbiology* (pp. 43–8), *Great Experiments in Biology* (pp. 110–18) and *A History of Biology* (pp. 443–52).

A very curious historical outcome of the debate is that in fact the observations of both were correct. Pouchet's results were not due to faulty technique as Pasteur suggested. If Pouchet had not withdrawn from the contest, the French Academy of Science might well have been placed in a difficult situation.

This point is taken up by Conant in *Harvard Case Histories*:

Three points of special interest stand out from this historic mistake of Pasteur's in the 1860s. The first is how easy it is to assume that two sets of experimental conditions are effectively identical in spite of their obvious differences. Why did not Pasteur try Pouchet's experiments with hay infusion? Probably because he was convinced that all fermentable materials were essentially the same (yeast water with sugar, vegetable infusion, mutton broth, what did it matter?). The second point to be noted is the practical consequence of the demonstration that the spores of certain bacteria were heat-resistant. There soon developed the practice of sterilising at temperatures above the boiling point of water, by using an autoclave (that is, a modified pressure cooker). This is routine procedure in bacteriology today.

The third point is more philosophical in nature. Every day, probably someone in some laboratory or hospital finds a supposedly sterile medium that shows growths. But no one in his or her right mind would suggest that this was due to spontaneous generation. Rather, it would be assumed that some error had been made, that the flasks or media had not been sufficiently sterilised. And no one today would take the time and trouble to study such cases of apparent spontaneous generation. The impossibility of the kind of spontaneous generation that Pouchet and Bastian believed in has thus become an accepted premise among scientists. Why is this so? Are Pasteur's and Tyndall's experiments so convincing? Can one safely generalise so widely even from a mass of negative evidence, particularly if one must, apparently arbitrarily, throw out the positive cases as being due to 'faulty manipulations'? (vol. 2, pp. 442–3)

PANSPERMISM (GERM THEORY)

The doctrine of spontaneous generation was replaced by the theory of panspermism. The problems involved in the full acceptance of the germ theory is discussed by J. K. Crellin in 'Airborne Particles and the Germ Theory: 1860–1880', *Annals of Science*, Vol. 22, No. 1 (March 1966):

There are two main reasons why the controversy between the supporters of spontaneous generation and those of the germ theory lasted for many years after Pasteur's well-known and widely accepted criticism of spontaneous generation during the early 1860s. They are, first, the slow realisation of the thermal resistance of some micro-organisms, or rather of their spores, and secondly, the difficulty of obtaining direct evidence that micro-organisms were present in the atmosphere. It is the purpose of this paper to consider the second of these reasons and to examine the germ theory (panspermism) as it was understood in the two decades between 1860 and 1880. The theory, though incorrect in detail, eclipsed the doctrine of spontaneous generation and, it will be seen, slowly but directly led to the germ theory still accepted today. The correct germ theory was accepted almost twenty years after Pasteur's studies, published in the early 1860s, which are usually considered to have established the theory.

MECHANISM AND MATERIALISM

The approach of these two schools of thought was somewhat different. The mechanists believed that life originated from inorganic matter in accordance with natural laws. This view had to overcome the objection that only life could produce organic matter. They overcame this difficulty by postulating a macro-molecule, that is, large organic molecules with carbon skeletons, without which protoplasm cannot exist. So their theory depends on chance combination of inorganic molecules to produce these macro-molecules.

The materialists do not consider that organic matter can be produced only by living things. Their belief, therefore, is that organic material can be produced abiologically, these steps taking place before the origin of life. While the mechanists believed in chance formation, the materialists believed that in the evolution of increasing complexity, life was, as it were, inevitable. One can trace this idea in the writings of many; names to include are those of Engels, Spencer, T. H. Huxley, Mitchel and Schäfer. Some of these views were given by Schäfer in 1912 in his presidential address to the British Association for Advancement of Science:

> Setting aside as devoid of scientific foundation the idea of immediate supernatural intervention in the first production of life, we are not only justified in believing, but compelled to believe, that living matter must have owed its origin to causes similar in character to those which have been instrumental in producing all other forms of matter in the universe, in other words, to a process of gradual evolution.

WHAT IS LIFE?

In discussing the problem of life, one is sooner or later faced with the question of just what it is. How can one define life? In this connection, one feels in complete agreement with Pauling, when he stated:

In connection with the origin of life, I should like to say that it is sometimes easier to study the subject than to define it.

The problem still remains, and modern work on virus particles has not made the problem any easier. There are two views to be considered. In the first place, we have Horowitz' minimum properties for a living system: mutability, self-duplication and heterocatalysis (see Keosian, pp. 6, 62 and 70). Horowitz does point out that these criteria of life can be satisfied by molecules of DNA, so his definition of life becomes 'individual molecules in a polymolecular environment'. In the second place, we have the views of Madison:

Life is anything perceivable by an observer of a group of chemical systems, in which free energy is released as a part of the reactions of one or more of the systems and in which some of the free energy is used in reactions of one or more of the remaining systems. ('The Organism and its Origin', *Evolution,* 7, 211 (1953))

THE IMPORTANCE OF GEOLOGY

Geology remained longer than any of the other sciences under the influence of philosophy and theology. It is not until the late eighteenth century that we find true scientific thinking in geology. While today fossils are considered to provide one of the most important direct evidences for evolution, this has not always been the case. Early ideas on fossils are well brought out by H. T. Pledge in *Science since 1500* :

Leonardo da Vinci, Giordano Bruno, Stevinus, and other men of the Renaissance, with their vivid freedom from tradition, at once reached such common-sense conclusions as, that the rivers had sculptured the valleys, that some rocks had once been molten, that the sea, which is (Bruno) always near active volcanoes, must have some connection with them, that the land and the sea may not always have been in their present positions. But the clarity of these early views was soon lost. It was unsupported by work in the field, and put side by side with pre-scientific ideas. Fossils, for example, were Sports of Nature played on mankind, or the earth's imperfect successes in its efforts at the (spontaneous) generation of life. (p. 91)

Hooke used for fossils the metaphor of a coin. He also suggested extinct species, as opposed merely to dead individuals, to account for these relics. While this last view was strongly opposed, it could itself issue in fallacy and speculation. The Flood (asserted the 'Diluvialists') had destroyed previous creations, of which the fossils were relics. When the eighteenth century revealed several very different series of fossils, a whole series of 'Catastrophes' was asserted, without evidence, by the school of that name. (p. 92)

After each such catastrophe, a fresh creation of life was essential. This idea, therefore, denies any evolutionary connection between successive series of fossils.

UNIFORMITARIANISM VERSUS CATASTROPHISM

At the end of the eighteenth century, theories of the earth were developed in a more scientific way. The first of these was that of James Hutton (1726–97), whose *Theory of the Earth* was published in 1795. It was he who first suggested that the geological forces operating today are the same as those in the past. His work was developed by John Playfair in *Illustrations of the Huttonian Theory of the Earth* (1802), and by William Smith in *Stratigraphical System of Organised Fossils* (1817). However, the greatest influence of the period was exercised by Sir Charles Lyell (1797–1875). He, more than anyone else, was responsible for the final downfall of catastrophism. His major work was *Principles of Geology* (1830–3). He developed further the idea of the uniformity of geological forces, and also the idea that the earth was old enough for the process of evolution to have taken place. His influence on geologists in this country was possibly greater than on the Continent. Perhaps his greatest influence, however, was on biologists, Darwin in particular. The acceptance of uniformitarianism therefore had a profound influence on the development of biology. Catastrophism may well be regarded as the last stand against the idea of evolution.

Ideas die hard. Even in the mid-nineteenth century there was a revival of catastrophic ideas, particularly by Jean Louis Rodolphe Agassiz (1807–73). His views are expressed by Isaac Asimov in his book *A Short History of Biology*:

The Ice Age (now known to have existed as four separate Ice Ages within the last half-million years or so) was good evidence to the effect that the extreme uniformitarianism of Hutton and Lyell was not justified. These were catastrophes, after all. To be sure, these were not as sudden, as shattering and as fatal as Cuvier's theories demanded, but they existed. Between his own Cuvier-like feeling and his natural piety, Agassiz found that he was incapable of accepting the Darwinian theory. (p. 64)

(See also the chapter 'The Earth Acquires a History' in *The Discovery of Time*.)

EARLY FORMATION OF LIFE

Modern speculation on the origin of life was stimulated particularly by the work of J. B. S. Haldane and A. I. Oparin. We can trace the extent of their ideas in Oparin's own words:

... This need not lead us to the conclusion that there is an absolute and fundamental difference between a living organism and lifeless matter. (p. 246)

... At first there were the simple solutions of organic substances, whose behaviour was governed by the properties of their component atoms and the arrangement of these atoms in the molecular structure. But gradually as a result of growth and increased complexity of molecules, new properties have come into being and a new colloidal chemical order was imposed upon the more simple organic chemical relations. (p. 248)

At the beginning there was considerable theorising about the type of molecule in existence on the primitive earth, and the reactions that later took place. In recent years, particularly in the United States, chemists and biologists have been actively investigating the formation of organic compounds and protein molecules. The names of M. Calvin, S. Miller, H. Urey and S. Fox are associated with this work.

Students will no doubt find chapter 5 of *Biological Science: Molecules to Man* ('The Forerunners of Life') informative and stimulating. This could well provide useful background to class discussion. In investigation 14, in this book, the formation of coacervates (organised droplets) is studied. This laboratory work could contribute to the understanding of this section. Since coacervates exhibit some of the properties of living things, their formation is one very important step in the origin of life hypothesis. In this investigation solutions of gelatin and gum arabic are mixed together, forming coacervates.

In *The Origin of Life* J. Keosian considers some of the problems:

It was not entirely unexpected that controversies should arise. Theories on the origin of life have many facets, all of them in dispute in some measure, as for instance, the time and manner of formation of the earth and its atmosphere, the composition of the primitive atmosphere, the kinds of organic compounds that would form such an atmosphere, the mechanism of further increase in complexity of organic compounds, the nature and magnitude of the energy required to drive the chemical reactions, and the mechanism of formation of the first living things.

No one set of conditions for the primitive earth and its atmosphere has been universally accepted. Experimenters proceed on the assumptions that they consider most likely. These assumptions vary all the way from primitive earth conditions similar to the present, to conditions that picture the primitive earth as a hot body with a seething turbulent atmosphere of gases that would be noxious to any present living thing. However, the proposed starting conditions are merely a point of departure for specific ideas on organic chemical synthesis. The surprising realisation that is emerging is that organic compounds of much the same variety, if not abundance, can be expected not from one, but from quite a number of different starting mixtures under the influence of many different energy types. (p. 87)

NEOBIOGENESIS

One can ask whether these theories concerning the origin of life are in fact resurrecting the old controversy of spontaneous generation. It is true that in both neobiogenesis and spontaneous generation life originates from inorganic material. But the likeness stops there. Neobiogenesists believe in the formation and separation of macromolecules before the evolution of living systems. However, many of the objections to spontaneous generation have been applied to neobiogenesis. The acceptance of neobiogenesis may have far-reaching consequences in biology. One assumption is that life at a very primitive level continued to be formed even after the appearance of the first living things. This would in turn mean that we would have to re-examine the interrelatedness that we assume exists between living organisms. Not only may the plant and animal kingdoms have had separate origins, but possibly also different phyla. This subject is considered by Keosian in *The Origin of Life* 'Speculations', pp. 92–112).

Speculation

The following passage from *Early Evolution of Life* by R. A. Young and C. Ponnamperuma is worth considering in discussion:

Biology, unlike most of the physical sciences, is lacking in universal principles because of its restriction to the planet earth. Perhaps the nearest thing to a universal biological principle is that of Darwin's theory of evolution through the natural selection of random hereditary fluctuations. A discovery of extra-terrestrial life would determine the universality of this and perhaps other concepts. [p. 21; on pp. 22 and 23 the subject of extra-terrestrial life is discussed.]

III. 3 Cytology

In *A History of Cytology,* A. Hughes begins his preface:

An account of so small a fragment of the sum total of human experience as the development of microscopical research can hardly hope to gratify the general historian to any appreciable extent. It is of no special significance, for instance, that the discovery of Brownian motion preceded the emancipation of Roman Catholics in England by a single year. It is certainly true that social history now recognises the importance of scientific discovery as a factor in economic progress, but in this tends to concentrate on such knowledge 'as hath a tendency for use', as Robert Boyle put it. Thus pure research is overshadowed by technology in the story of past centuries as in the expanding universities of the present day. (p. ix)

One must consider, however, how the invention of the microscope in the seventeenth century opened up a totally new world, a world so small that it had hitherto been invisible. Researches with this new invention opened the way to many avenues of thought in general biology. One can consider its effect on the development of the theories of evolution and inheritance, and in the applied field of medicine—the theory and control of disease.

If a biologist were asked today to explain the nature of living matter, he would in all probability answer by stating a number of basic principles. One can consider those put forward in Hughes:

(1) Living matter, in all but the smallest organisms, is nearly always divided into units, which we call cells. The great majority of cells contain a special organ, the nucleus.
(2) Cells originate only from pre-existing cells.
(3) When the cell divides, the nucleus divides first and the two nuclei which are formed are exact replicas of the original one. (p. 1)

These indeed could be regarded as being part of the content of any present-day elementary biology syllabus. However, when the *Origin of Species* was published just over a century ago, the last two principles were totally unknown to biologists.

Though the cell has been accepted as the unit of life for only a little over a century, the history of cytology can be traced to the mid-seventeenth century. The first cell to be observed was probably the human blood cell. We have the reports of Pierre Borel in 1656, and some years later, in 1661, Marcello Malpighi refers to 'utricles', 'globules' and 'saccules'. It is in 1665 that we find the first mention of the term 'cell'. It was introduced by Robert Hooke (1635–1703) in *Micrographia*. Hooke described the cells of cork:

I could exceedingly plainly perceive it to be all perforated and porous, much like a Honey-comb, but that the pores of it were not regular; yet it was not unlike a Honey-comb in these particulars.

He described and examined other plants:

Nor is this kind of texture peculiar to cork only, for upon examination with my microscope, I have found that the pith of the Elder or almost any other tree, the inner pith of the Cany hollow stalks of several other vegetables: as of Fennel, Carrets, Daucus, Bur-docks, Teasels, Fearn, some kind of Reeds etc. have much such a kind of Schematisme, as I have lately shown that of cork.

Following the observations of Hooke, there were the well-illustrated volumes on the microscopic structure of plants produced by Nehemiah Grew (1641–1742). Also, in 1674, animals we now classify as protozoa were observed by Antoni van Leeuwenhoek (1632–1723). From this time until the beginning of the nineteenth century the study of cytology remained more or less at a standstill.

EARLY NINETEENTH-CENTURY CYTOLOGY

Early in the nineteenth century there was a renewed interest in the investigation of cell structure. Two names stand out during this period. The first is Lorenz Oken (1779–1851), who stated that animals and plants consisted of vesicles or cells. Singer, in *A History of Biology,* discusses the meaning of the term 'cell' as used by Oken in a passage (pp. 220–3) in *Die Zeugung* (Generation, 1805). Singer states:

The passage is an illustration of an interesting feature in the history of science. We often encounter, in works of exceptional ability, the enunciation of truths that are beyond the comprehension of their author. The presentation of premature views is naturally liable to be confused. The vocabulary of an age reflects the interests and beliefs of that age. The character of the scientific ideas is reflected in the character of contemporary terminology. With an inevitably changing vocabulary, we must beware of reading into words used in older works meanings that are too modern for such connotation. Yet there are writers with whom such caution may at times be misplaced. Oken can and does use the word cell in the modern sense. His works were widely read by his contemporaries, and it is more than probable that he inseminated the minds of the recognised founders of the cell doctrine. (pp. 332–3)

The second name is that of Robert Brown (1773–1858), who in 1831 recognised the nucleus in the plant cell.

Hughes suggests that from around 1830 onwards the discoveries can be classified under three headings:

(1) Recognition of nuclei within the cells of plants, and in the earliest stages of animal development.
(2) The discovery of cells within the tissues of animals.
(3) Recognition of the nature of the primary living substance and its first designation by the word 'protoplasm'. (p. 34)

The realisation that the cells of all organisms are similar came very gradually, and towards the mid-nineteenth century we find the first attempt to establish a cell theory. This was the work of Mathias Jakob Schleiden (1804–81) and Theodor Schwann (1810–82).

We see in Schwann's work a projection of Descartes' idea of a century earlier (the organism as a machine-using energy), since Schwann stressed, in addition to the *morphology* of the cell, that *chemical changes* take place within it. This he called *metabolism*. Although this recognition of the chemical aspects of the cell came at this relatively early time, it had no influence on nineteenth-century cytology, which remained predominantly morphological. Indeed, the co-relation of chemical activity with the ultra-structure of the cell is a very recent innovation.

Schleiden stressed the importance of the study of development as the best way of approaching the study of plant structure. He had a considerable influence on Hofmeister, Nägeli and Zeiss, and particularly on Schwann. We can see how this developed in Schwann's own words:

One day, when I was dining with M. Schleiden, this illustrious botanist pointed out to me the important role that the nucleus plays in the development of plant cells. I at once recalled having seen a similar organ in the cells of the notochord, and in the same instant I grasped the extreme importance that my discovery would have if I succeeded in showing that this nucleus plays the same role in the cells of the notochord as does the nucleus of plants in the development of plant cells. (*Frédéricq,* 1884)

Schwann's general statement on the cellular structure of plant and animal cells has been expressed in the following way by Singer in *A History of Biology*:

(a) The entire animal or plant is composed either of cells or of substances thrown off by cells.
(b) The cells have a life that is to some extent their own.
(c) This individual life of all the cells is subject to that of the organism as a whole. (pp. 338–9)

Schwann, however, failed to correct a false idea Schleiden had introduced into the cell theory, namely that new cells are derived by a process of budding from the nucleus. The cell bud he called cytoplast. This mistake is well documented in *Architecture*:

Having originated in botany rather than zoology, it continued to place undue importance on the wall surrounding the cell. The nucleus was treated as a temporary feature, which had a part to play only during the formation of the cell wall, and the theory effectively ignored the remaining (apparently fluid) contents of the cell—i.e. the cytoplasm. (p. 399)

At this stage it is worth considering two quotations from Hughes:

The observation was correct, but the theory, alas, was utterly wrong. And unhappily, the young Theodore Schwann unquestioningly adopted the whole of it from his 'illustrious' senior. The history of the cell doctrine for the next fifty years is mainly concerned with the results of this fatal mistake. (p. 38)

And further:

Had Purkinje's contribution to this field not been overshadowed by that of Schwann, the mistaken views of Schwann might have exercised less influence in subsequent years. (p. 39)

Criticism of the cell theory of Schleiden and Schwann came from Hugo von Mohl (1805–72). It was he who independently reintroduced the term 'protoplasm', a term previously used by

Purkinje. In his cytological studies von Mohl had observed that the cell material contracts away from the cell wall when material is left in spirit for some time. He also observed that new cells derived from the parent cell were smaller, and only later expanded to the original size of the parent cell. Furthermore this process could not take place in cells containing granules such as chlorophyll. These observations led him not only to refute the cell theory of Schleiden and Schwann, but to comment that Schleiden could not have observed the division of cells.

The confusion as to the origin of cells no doubt hindered the development of embryology and pathology during this time.

CELL DIVISION

New methods of staining introduced in the last quarter of the nineteenth century contributed in no small way to the understanding of cell division. Major work in this field was done by Eduard Strasburger (1844–1912) and Walther Flemming (1843–1915). The former was concerned with the division of plant cells, and the latter with the division of animal cells. Their work demonstrated that the process of cell division was similar in animal and plant cells, and that the process was controlled by the nucleus.

These accounts included reference to a resting nucleus, chromosomes, splitting of the chromosomes and the formation of daughter nuclei. Flemming considered from the beginning that the splitting was longitudinal while Strasburger at first considered the division to be transverse.

The term 'mitosis' was introduced by Flemming in 1882, and in the same year Strasburger introduced the terms cytoplasm and nucleoplasm, these terms having the meaning we give them today. Strasburger also introduced the terms 'prophase', 'metaphase' and 'anaphase' for the first three stages in the process.

Further developments took place a few years later. Theodor Boveri (1862–1915) in 1887–8, and Oscar Hertwig (1849–1922) independently two years later, described the process of 'reduction division' in animals. Guignard, in 1899, was one of the first to demonstrate the process in plants.

References to this development in biology can be found in Singer's *A History of Biology* (pp. 345–8 and 534–40). The process of mitosis and meiosis may not be familiar to those students not taking biology. There are two excellent films, *Mitosis* and *Meiosis*, which could be used here. Also there are 8 mm film loops developed by both Nuffield and B.S.C.S. Biology Teaching Projects: these would be most suitable teaching aids.

THE NUCLEUS AND HEREDITY

As might be expected, the study of the cell and the nucleus has been of special interest to those studying the process of heredity. One of the earliest was Charles Darwin (1805–82). He put forward the hypothesis that body cells give off 'gemmules', and that these have the property of developing into new cells. In the gonads these gemmules produce germ cells. The gonads have gemmules not only from all parts of the body but also from all stages of development, including even past generations, the idea being that a large number of these gemmules could be passively transmitted through each generation. Darwin's 'pangenesis' became very closely linked with Lamarckian inheritance.

The next major hypothesis was that suggested by August Weismann (1834–1914). His approach was different from that of Darwin. To quote Hughes:

> Whereas Darwin had been primarily concerned with suggesting a mechanism whereby organisms could change in character during the long course of many successive generations, Weismann's view may be said to begin with the necessity of accounting for the stability of reproduction of living creatures. How is this pattern of organic design formed afresh in each generation, and a virtual replica of the parent so produced? (p. 81)

One can consider Weismann's ideas from his own words in *Essays upon Heredity* (authorised translation, Oxford, 1889):

> Now if it is impossible for the germ cell to be as it were an extract of the whole body, and for all the cells of the organism to despatch small particles to the germ cells, from which the latter derive their power of heredity; then there remains, as it seems to me, only two other possible theories as to the origin of germ cells. Either the substance of the parent germ cell is capable of undergoing a series of changes which, after the building up of a new individual, leads back again to identical germ cells, or the germ cells are not derived at all, as far as their essential and characteristic substance is concerned, from the body of the individual, but they are derived directly from the parent germ cell. I believe that

the latter view is the true one. . . . I propose to call it the theory of 'The Continuity of the Germ Plasma', for it is founded upon the idea that heredity is brought about by the transmission from one generation to another of a substance with a definite chemical and, above all, molecular constitution. (p. 167)

By 1892, Weismann's 'heredity substance' could be identified with the chromosomes. However, this theory cannot be regarded as being epigenetic, since he misunderstood the nature of reduction division. At the turn of the century, research became concerned more with the study of the individual chromosomes than with the nucleus as a whole. Since the rediscovery of Mendel's work, it has been in the main concerned with the concept of the gene.

THE CYTOPLASM

The successful history of cytoplasmic cytology is extremely short: real progress in the study of the cytoplasm has had to wait for modern technical advances. Two advances in particular have made this study possible: first the development of histo-chemical methods, and second, the invention of the electron microscope (see *Architecture,* pp. 409–10).

The first cells to be examined for their ultra-structure were the nerve cells, by Treviranus in 1816, Ehrenberg in 1833 and 1836, and Romak in 1844. From this time onward there were numerous accounts of the net-like structure of the cytoplasm. These structures were interpreted in different ways, for instance the 'spongiplasm' and 'hyaloplasm' of Franz von Leydig (1821–1905), representing the solid framework enclosing a fluid.

The problems of the early cytoplasmic cytologists were not very different from those of today. The main problem is the degree of distortion and disturbance caused by the various techniques. One of the earliest to point this out was W. B. Hardy of Cambridge. A good example of this is afforded by the 'Golgi-apparatus'. Discovered in 1898 by Camillo Golgi (1844–1926), its role, if not its existence, has been far from clear even up to recent times.

One of the earliest cytoplasmic structures to be shown were the 'mitochondria', named by Benda, who first observed them in 1880 as spiral-shaped structures in the middle piece of the spermatozoa of a mouse. The association of respiratory enzymes with these structures came much later, the main evidence coming from the work of Bersley and Hooer in 1934. They were able to separate large quantities of mitochondria from liver cells. (See ch. 5 in Hughes' *A History of the Study of Cytoplasm.*)

AGEING AND DEATH

There are many questions still to be answered, one of which has special social significance, the problem of ageing and death. The cell, like a complete organism, has its homoeostatic mechanism which regulates the many chemical reactions that take place within it. There is no doubt that a mechanism exists to control the processes of growth and reproduction. Is there also a mechanism that brings the life span to an end? If so, where is the location of this biological clock? Could this be set, so as to maintain a vigorous species? The question may also be asked whether it is desirable to increase the life span? Reference can be made to the chapter 'Ageing and Death' in Butler's *Inside the Living Cell.*

MOLECULES AND MEMORY

Research in cytology, more than in any other branch of biology, tends to provide additional information with which to study, and sometimes even solve, problems in other fields. This has been very true in the development of genetics. Now we find research in biochemical cytology aiding the understanding of brain function. Evidence is accumulating that alteration in the rate of protein and RNA synthesis within the cells of the brain may provide a theory based on biochemical machinery for memory.

Reviews of experimental evidence in this field can be found in: 'The Molecules of Memory', by Steven Rose, *Science Journal,* May 1967, and 'Memory and Protein Synthesis', by Bernard W. Agranoff, *Scientific American,* June 1967.

III. 4 Genetics

Bateson had this to say of heredity:

An exact determination of the laws of heredity will probably work more change in man's outlook on the world, and in his power over nature, than any other advance in natural knowledge that can be clearly seen. ('Problems of Heredity and their Solution', *J. Roy. Hort. Soc.*, **25**, 54–61, 1900)

PRE-MENDEL

There was no lack of interest in heredity before the time of Mendel. Aristotle was interested in heredity, and Hippocrates, even earlier, had put forward an idea similar in many respects to Darwin's idea of pangenesis (see Sturtevant, p. 1). There is also a record of artificial pollination of a date palm around 2400 B.C. This must indeed suggest a kind of deliberate selection of desired characteristics. There are also many references to artificial pollination (hybridisation) by Joseph Gottlieb Koelreuter (1733–1806). Towards the end of the eighteenth century there is the work of Bakewell, Collings and Bates on the improvements of British breeds of sheep and cattle. During the first quarter of the nineteenth century there are the observations of Goss (1822), Seaton (1822) and Knight (1823) on peas, all three observing segregation of characteristics, but failing to detect the numerical ratios. (See Sturtevant, ch. 1, 'Before Mendel'.)

1866–1900

In 1866, Gregor Mendel (1822–84) published his paper on 'Experiments in Plant Hybridisation'—the paper was presented to the Brünn Natural History Society in 1865. For thirty-four years (that is between its publication and rediscovery), Mendel's work had no effect on the development of genetics. This period was dominated to a considerable degree by the development of the Darwinian theory of evolution, and all its implications. There were also very significant advances in the field of cytology, advances that not only helped to make Mendel's results acceptable, but also provided the foundations for the very rapid advance made in this new field after the rediscovery of Mendel's paper.

Chromosomes were described in 1875 by Eduard Strasburger (1844–1912), and in the same year Oscar Hertwig (1849–1922), showed that the process of fertilistion involved the fusion of two parental nuclei. The process of cell division, including the longitudinal splitting of the chromosomes, was observed by Walther Flemming (1843–1915) during the period 1879–83, and soon afterwards he observed the daughter chromatids passing to opposite poles of the cell. The work of these early cytologists and that of August Weismann established that the nucleus was the basis of inheritance. Weismann had predicted the occurrence of meiosis in the year 1887 (see Sturtevant, pp. 17–20).

There were also theories of inheritance, in this Darwinian epoch, based on a belief in blended inheritance. Darwin himself put forward the idea of 'pangenesis': the sex glands received particles from each organ of the body, modifications in these organs being represented in the particles and passed on to the next generation. Weismann's theory of 'germ plasma' had more influence. This theory was based on the assumption that the resemblance of offspring to parent is simply the result of both being derived from the same substance. In fact, as Samuel Butler (1835–1902) expressed it, 'a hen is only an egg's way of producing another egg'.

Another attempt at a theory of inheritance was the 'law of ancestral inheritance' put forward by Francis Galton (1822–1911). Galton is well known for his interest in the subject and for coining the term 'eugenics', and for his contribution to statistics. According to Galton, an offspring received half its heritage from both parents, half the remainder from the four grandparents and so on.

All these theories have a common failing. They are not particulate in the modern sense, that is, not retaining their properties and identity.

Hugo de Vries (1848–1935) developed a theory of heredity that differed from the theory of Weismann. De Vries recognised that there were two parts to Darwin's pangenesis hypothesis. He accepted the first part, the idea of persistent heredity units, but rejected the idea of somatic influence. He called his units 'pangens'. These units were particulate, found in the nucleus, and they could multiply. They were extraordinarily constant, but still capable of variation.

Towards the end of the century there was also considerable dissatisfaction with the view that evolution could be explained in terms of selection. William Bateson (1861–1926), de Vries and others placed much greater emphasis on discontinuous variation.

The rediscovery of Mendel's work was made by de Vries, Carl Correns (1864–1933) and Erich von Tschermak (1871–1962). All three had carried out work confirming the principles established by Mendel thirty-four years earlier (see Sturtevant, pp. 25–32).

MENDEL

Mendel's paper, *Experiments in Plant Hybridisation,* was published in 1866. The fact that it had no impact for thirty-four years must represent one of the most remarkable cases of discontinuity of discovery. It is even more remarkable considering that his theory had not been anticipated, and considering also the extent to which Mendel developed its theoretical consequences.

Many reasons have been put forward to account for this. These range from the fact that Mendel's paper was first published in an obscure journal, to the fact that contemporary scientists simply did not recognise its importance. At this stage perhaps it is worth considering the views of two geneticists. In *The Gene—A Critical History,* Carlson puts forward his idea:

I think it would have been possible for Mendelism to have had its start in 1865, and not 1900. The failure was due to Mendel himself. Unlike Bateson, Mendel did not debate with his critics. He did not stir up a storm with Nägeli. Unlike Bateson, Mendel did not widely publicise his results. Mendel as far as it is known did not send a reprint of his work to Darwin, although he was familiar with the importance of Darwin's contributions. (p. 245)

A similar view is put forward by Sturtevant in *A History of Genetics.* After stating that Mendelism is one example of how important it is that the time be ripe for a particular theory to be accepted, he goes on:

It is true that the paper was ahead of its time, but it was not difficult to understand, and it seems unlikely that it would have remained unappreciated for long if it had appeared in a less obscure journal, or if Mendel himself had publicised the further cases he reported in his letters to Nägeli. It must be remembered that Nägeli's failure to appreciate the paper in 1866 can be matched by Pearson's failure in 1904. Both were outstanding men, and both were actively studying heredity, but to both of them Mendel's results appeared as trivial cases involving a few superficial characters, obviously neither useful nor illuminating for any general theory of inheritance. It does not follow that no biologist was likely to have appreciated the paper if he had seen it before 1900. I have suggested above that Galton, for one, would very likely have done so. (p. 133)

These quotations bring out a very important factor which cannot be ignored, namely the place of the scientist in discovery. It would be an interesting exercise to compare and analyse the personalities of Harvey, Darwin and Mendel. Also one should consider the extent to which a scientist's personality and reputation affects the acceptance of his pronouncements.

Even after rediscovery, Mendel's work and ideas did not go unopposed. The main opposition came from supporters of the Galtonian 'law of ancestral inheritance', whose main exponents were Weldon and Pearson. According to Galton, inheritance depended on at least half a dozen generations: an individual inherited one quarter of his or her characteristics from each parent, one sixteenth from each grandparent, one sixty-fourth from each great-grandparent and so on. According to Mendel, inheritance depended on the parents alone. The controversy resulted in an anti-Mendelian stand at the beginning of this century, and had the effect of delaying the application of statistical methods to the study of genetics. However, one must realise that the controversy was not Galton's fault, nor of his making.

Galton also appears connected with two other aspects of inheritance. He carried out experiments to test Darwin's theory of pangenesis. However, after performing extensive blood transfusions between different strains of rabbits, he found no effects on the offspring. His other work was concerned with human genetics. He published *Hereditary Genius* in 1869, and *Natural Inheritance* in 1889. In his earlier work he studied 300 families which had produced one or more men of distinction. This point is discussed by Dobzhansky (pp. 77–9). He states:

Clearly, eminence 'runs in families'. But does it follow that possession of a certain genetic endowment is necessary or sufficient to attain eminence? Surely having influential relatives is helpful, even in societies with class barriers less rigid than those of nineteenth-century England. Galton was not oblivious of this possibility. But he dismissed it because he defined the genetic endowment of eminent men as that which, 'when left to itself, will, urged by an inherent stimulus, climb the path that leads to eminence, and has strength to reach the summit—one which, if hindered or thwarted will fret and strive until the hindrance is overcome, and it is again free to follow its labour-loving instinct'.

This sounds perilously close to circular reasoning. Galton's clinching argument seems whimsical at present, but it was taken seriously a century ago. Class barriers are less rigid in the United States than in England; it should be easier to achieve eminence in the former than in the latter country; one might accordingly expect that the United States will produce more men of genius than England; in point of fact, the opposite is true; therefore, according to Galton, to become eminent one must inherit genes that are more frequent in the English than in the American population. (p. 78)

The work carried out by Galton has been repeated many times and his observations confirmed, but not his conclusions. Dobzhansky discusses 'genetic conditioning' in chapter 4, 'Equal but Dissimilar'. The film *Heredity in Man* can also be shown at this point. This film shows the application of Mendelian laws to human affairs. It shows the Phelps pedigree (good physique), the Godfrey pedigree (musical talent), the Terry pedigree (artistic talent) and also a pedigree showing members who are mentally defective.

Mendel's paper is relatively short and is not difficult to read. A translation appears in *Classic Papers in Genetics,* edited by J. A. Peters (Prentice-Hall). There are special points to notice, for they show extremely well the method of science: the planning of the experiment, the mathematical patterns involved, the importance of a large sample, and of controlling the pollination, and then allowing the F_1 to pollinate themselves. From his results Mendel could make predictions. One in the series of letters to Carl von Nägeli from Mendel can be read in *Great Experiments in Biology,* edited by Gabriel and Fogel (Prentice-Hall). This letter summarises much of Mendel's original experimental work.

One point that should be stressed is Mendel's confidence in his work and results. Two short quotations from 'Experiments in Plant Hybridisation' bring this out:

. . . among the numerous experiments made [here he is making reference to his predecessors], not one has been carried out to such an extent and in such a way as to make it possible to determine the number of different forms under which the offspring of hybrids appear, or to arrange these forms with certainty according to their separation generations, or definitely to ascertain their statistical relations.

. . . it is now clear that the hybrids form seeds having one or the other of the two differentiating characters and of these one-half develop again the hybrid form while the other half yields the plants which remain constant and receive the dominant or the recessive characters (respectively) in equal numbers.

The film *Gregor Mendel* or *Heredity in Animals* can be shown at this stage to demonstrate some of the Mendelian principles of heredity.

R. A. Fisher (1890–1962) expressed the view (1936) that Mendel may have anticipated the results before he started his experiments. Mendel's ratios, according to Fisher, are consistently closer to what is expected than is perhaps allowed by the sampling theory. Fisher's papers on this subject—'Marginal Comments on Mendel's Paper', and 'Has Mendel's work been Rediscovered?'—appear in a book with Mendel's paper. The book is published under the title of Mendel's paper, *Experiments in Plant Hybridisation.* This point is discussed at considerable length in Sturtevant's *A History of Genetics* (pp. 12–16). A table is also included of the F_2 results for pea crosses repeated by many investigators. In these it is interesting to note that the results of Tschermak show an even closer approach to the 3:1 ratio than those of Mendel.

Sturtevant's summary is as follows:

In summary, then, Fisher's analysis of Mendel's data must stand essentially as he stated it. There remains the question of how the data came to be as they are. There are at least three possibilities:

1. There may have been an unconscious tendency to classify somewhat doubtful individuals in such a way as to fit the expectation.

2. There may have been some families that seemed aberrant and that were omitted as being probably due to experimental error.

3. Some of the counts may have been made for him by students or assistants who were aware of his expectations, and wanted to please him.

None of these alternatives is wholly satisfactory, since they seem out of character, as judged by the whole tone of the paper. Perhaps the best answer—with which I think Fisher would have agreed—is that, after all, Mendel was right! (p. 16)

THE CONCEPT OF THE GENE

The study of the development of the concept of the gene—the 'factor' of Mendel was renamed 'gene' by W. J. Johannsen (1857–1929)—presents many difficulties. Some of these difficulties are listed below:

(a) In the first place the concept is still developing. One would expect this as the history of genetics is very short, from around 1900 to the present day.

(b) It reflects the complexities of modern science—teams of scientists working rather than individuals.

(c) Many of the scientists who have contributed to the development are not only alive, but may still be actively engaged in research.

(d) The picture is complicated by a tendency of many scientists to replace models or concepts rather than retract them.

(c) One may indeed question the possibility of making a definite historical assessment of such work at the present time.

To quote Carlson:

> It is the obligation of the historian of science, not the scientist, to reconstruct the development of concepts and to restore the fallen idols of previous generations to their original eminence. (p. 254)

The question is, when is such an obligation possible?

One can give a very long list of names of biologists concerned with the development of the gene. (Sturtevant gives a chronology of events on pp. 136–8.) It is possible, by reference to half a dozen or so papers, to trace the major events in the development of genetics:

T. H. Morgan, 'Sex-limited inheritance in drosophila', *Science* (**32**, 1910, pp. 120–2)

A. H. Sturtevant, 'The linear arrangement of six sex-linked factors in drosophila, as shown by their mode of association', *Journal of Experimental Zoology* (**14**, 1913, pp. 43–59)

H. J. Muller, 'Variation due to change in the individual gene', *American Naturalist* (**56**, 1922, pp. 32–50)

H. J. Muller, 'Artificial transmutation of the gene', *Science* (**66**, 1927, pp. 84–7)

G. W. Beadle and E. L. Tatum, 'Genetic control of biochemical reactions in neurospora', *National Academy of Sciences* (**27**, 1941, pp. 499–506)

O. T. Avery, C. M. MacLeod, and M. McCarty, 'Studies on the chemical nature of the substance inducing tranformation in pneumococcal types', *Journal of Experimental Medicine* (**79**, 1944, pp. 137–58)

J. D. Watson and F. H. C. Crick, 'Molecular Structure of Nucleic Acids', *Nature* (**171**, no. 4356, 1953, pp. 737–8)

All these appear in the collection *Classic Papers in Genetics.*

There are three main stages in the development of the 'gene' concept:

(a) A stage concerned with the location and mode of transmission of the gene, thereby confirming and developing Mendelian genetics.

(b) A stage concerned with the possibility of modification of the gene through mutation.

(c) A stage concerned with models of coding and regulation, forming part of the modern discipline of molecular biology.

There is no doubt that this third stage will extend perhaps for the rest of the present century. It is this last stage more than any other that is producing complications in terminology. Sturtevant brings this out:

The symbolism for genes has grown up on the basis of the older view, and it is still not clear what will be the most effective compromise. The older terms gene, allele, and locus are now in a fluid state so far as current usage is concerned, and several newer terms are in general use: cistron (Benzer) to denote an area that must be intact (that is, in the cis form) to produce the wild-type phenotype, and site (or recon of Benzer) to denote the smallest unit separable by recombination. It is still not clear what will be the most convenient system of terminology; probably it will depend on developments in the study of the genetic coding system. (p. 90)

The coding problem

The problem was clearly expressed by Bateson as far back as 1906 in *Progress in Genetic Research*:

But ever in our thoughts the question rings, what are these units that bring all this to pass, which in their orderly distribution decide so many and perhaps all of the attributes and faculties of each creature before it is launched into separate existence? Colour, shape, habit, power of resistance to disease, and many another property that might be named, have one by one been analysed and shown to be alike in the laws of their transmission, owing their excitation or extinction to such units or factors. Upon them the success or failure of every living thing depends. How the pack is shuffled and dealt we begin to perceive; but what are they—the cards? Wild and inscrutable the question sounds, but genetic research may answer it yet.

The answer or part of the answer came about fifty years later, in the DNA model of Watson and Crick (1953). DNA material was subjected to X-ray diffraction and from the results the three-dimensional structure of the molecule was worked out. This new technique is briefly described by Asimov (pp. 106–8). On page 142 of *Biological Science: Molecules to Man,* there is a photograph of X-ray diffraction of DNA. Also, in the film *A Light in Nature,* there is a short sequence showing the building up of a model of a large molecule. In a paper published in *Nature* ('Genetical implication of the structure of deoxyribonucleic acid') appeared the following statement:

It seems likely that the precise sequence of the bases is the code which carries the genetical information.

The bases are adenine, guanine, cytosine and thymine, non-randomly arranged. The sequence of these base pairs provides the code. Once the code had been recognised, the next problem was the understanding of the transmission process. During the period 1955–60, the relationship existing between genetic maps and polypeptide maps was established. This can be simply expressed in the following way:

$$\text{DNA}\rightarrow\text{RNA ('messenger RNA')}\rightarrow\text{Protein}$$

The first stage is a transcription of complementary maps, the protein formed being the analogue (see ch. 10 in Asimov's *The Messenger from the Nucleus*).

The next important problem is concerned with how many base combinations are necessary to provide a code for an amino acid. Evidence now seems to suggest that the code is a sequence of three nucleotides, or multiples of three. This replaces the overlapping idea first put forward by Gamow. In the case of four pairs of nucleotides, there are sixty-four triads possible. However, in the formation of proteins only twenty or so amino acids are involved. This leaves two possibilities, either not all the possibilities are represented, or it may be that different triads can represent the same amino acid (see Asimov, ch. 11, 'Breaking the Code').

TURNING BACK THE CLOCK—LYSENKOISM

Lamarckism (or neo-Lamarckism) has had its revivals in terms of 'Michurinism' and 'Lysenkoism'. Adherents to these ideas believe that the *environment* has a direct effect on and can alter the heredity material of an individual; this implies, in other words, the inheritance of acquired characteristics. The idea of pangenesis has also been re-invented by Lysenko, but his claims have not been substantiated. Galton tested the pangenesis hypothesis of Darwin, and found it wanting even at that time. Lysenkoism was strongly held during the life of Stalin, however, and although since his death the theory has had a steady decline, it has been periodically revived. This is an example of direct official political policy attempting to influence scientific thinking, the result being a restriction on the development of genetics in the USSR. In Lysenko's addresses we come across passages such as this:

> To us it is perfectly clear that the foundation principles of Mendelism–Morganism are false. They do not reflect the reality of living nature and are examples of metaphysics and idealism.

(See Carlson, pp. 95 and 255, for Muller's part in this controversy.)

THE SOCIAL IMPLICATIONS OF GENETICS

Of all the organisms for genetic study, man must be considered the most unsatisfactory. There are a number of reasons for this, including the time interval from one generation to another and the size of the family—even the largest families are too small to obtain scientificially acceptable ratios. Also, in the study of genetics it is essential to have test-matings, which in the case of humans cannot of course be carried out. The question that faces us then is this: to what extent are qualitative results obtained as a result of investigating one organism applicable when considering an entirely different organism? G. E. Allen makes this point in his paper 'General Education in the Sciences: The Role of history and philosophy' (C.U.E.B.S., December 1966):

This is important, for it emphasises that biological theories, like all scientific theories, have different degrees of certainty which are dependent upon the form of logic by which they were derived and the nature of the assumptions from which they arose. One of the chief assumptions of modern molecular biology is that the bio-chemical processes investigated in E. coli can be extrapolated to man and higher animals. What is the basis for this assumption? How certain are we of this, or any, extrapolation? The questions are important for they serve to undermine the commonly held view that all scientific information is somehow more certain than that of other fields.

Sturtevant makes the point with greater force. Here is a quotation from his paper 'Social Implications of Genetics to Man' (1954):

There is sufficient evidence that quantitative results obtained with one organism can be safely applied to a wholly different kind of organism.

He points out that certain findings have been widely confirmed as applicable to a variety of organisms; these include the effect of irradiation in producing mutations, and the fact that the great majority of these mutations are deleterious.

There is no doubt that the population already contains a high number of these deleterious mutations. The effect of irradiation of any kind is to add to this already high number. When considered in these terms, is there such a thing as being below a level which could be harmful? It is well worth discussing whether the expediency of obtaining short term benefits is worth the risk taken and whether sufficient consideration has been taken of the possible effect radiation may have on future generations.

The social implications of genetics would be incomplete without reference to eugenics. To this two approaches are possible. In the first, the negative approach, an attempt would be made to reduce or eliminate the number of defects (genetic) whether physical or mental. Problems immediately arise. There is not enough evidence for classifying a defect as genetical or environmental, in a large number of cases. Many serious diseases which are genetical in origin, e.g. Huntington's Chorea, do not appear in the early stages of life. Furthermore, what are we to do with recurrent mutations? In the second place there is the positive approach; here the number of 'better' individuals is increased. The problems here are equally great. The main question is who or whom will be set up to determine the basis on which selection is to be made? Even if such basis was determined, dominant characteristics may be dealt with relatively easily, but not the recessives.

The problems given are scientific ones, yet they serve to strengthen the ethical and moral objections when eugenics is applied to man.

(See Dobzhansky, pp. 13–15 and 322–5; also 'Radiation and Human Mutation' by H. J. Muller and *Scientific American*, November 1955, offprint no. 29.)

III. 5 Evolution

THE PROBLEMS OF BIOLOGY

Biologists have always been presented, indeed haunted, with two major problems: the classification of living things, and the definition of the term 'species'. These two problems are in fact closely related (see *The Discovery of Time*, pp. 172–7).

Classification

Singer, in *A History of Biology*, gives a very good account of the development of the various classification systems. What follows is therefore only a short sketch of some of the more important ideas.

Successful attempts at classification were made by the Greeks. The animal classification of Aristotle (384–322 B.C.)—Singer, pp. 42–3—and the plant classification of Theophrastus (380–287 B.C.)—Singer, pp. 44–53—deserve special attention. There is nothing approaching the excellence of Aristotle's work until the time of Lamarck (1744–1829). The increase in the number of plants and animals described resulted in many attempts at systematic grouping from the seventeenth century onwards. On the botanical side there is the work of John Ray (1627–1705)—Singer, p. 185—and Karl Linnaeus (1707–78)—Singer, pp. 191–2. These two may be regarded as the founders of modern systematic biology. It was Ray who introduced the system, later used by Linnaeus, of dividing and sub-dividing groups. It is interesting to note that Ray described some 10,600 plants, while Linnaeus' list included 70,000. Linnaeus studied the sexual structures of plants, and it was on this informajon that he based his system of classification in *Systema Naturae* (1734). He also attempted a classification of animals.

The important aspect of Linnean classification is that he grouped plants and animals into species; species into genera; genera into orders and orders into classes. He also introduced the system of 'binomial nomenclature', and his *Homo sapiens,* representing man, is still in use today.

The classification of Linnaeus has been revised many times, but many of his genera and species still remain. His classification had a considerable influence on the development of biology. Linnaeus also had very definite views on the nature of species. All species had been separately created, and all were maintained by divine providence. A species to him was constant, unchangeable (that is, in modern terms, not capable of mutation). One cannot therefore expect to see any attempt to show relationship or interrelationships in his classification, though in his genus *Homo* he included modern man, man the cave dweller, and the orang-utang. His classification must not be regarded as anything more than an artificial though convenient method of classification. The idea that a species could become extinct would have been anathema.

During the eighteenth and early nineteenth centuries considerable work was done in the field of classification. Of special note is the work of Augustin Pyrame de Candolle (Singer, p. 198), Georges Cuvier, and Jean Baptiste de Monet Lamarck (Singer, p. 199). There is a contrast between the classification of Lamarck and that of Cuvier, in that the former has a ladder or scale form. Lamarck was much influenced by Deist philosophy, a philosophy that influenced so much of biological thought in the eighteenth century. Lovejoy discusses the chain of being:

Thus the theory of the Chain of Being, purely speculative and traditional though it was, had upon natural history in this period an effect somewhat similar to that which the table of elements and their atomic weights has had upon chemical research in the past half-century. [*The Great Chain of Being,* p. 232; see also *The Discovery of Time,* p. 96.]

In Lamarck's classification we can see the conflict between the linear scheme of classification, as nature intended (ascending scale of perfection), and the observed facts, producing a somewhat branching scheme. In this is found the reason for Lamarck's theory of evolution, a theory to account for the deviations from a linear scheme. There is a marked change in the idea of constancy of species; in Lamarck's theory they do change, but they change to another species, they do not become extinct.

The term 'species'

From the nineteenth century onwards the concept of the species became more complicated, since it became involved in the origin of species. Quoting from Dobzhansky's *Mankind Evolving:*

. . . Can we define species? If we require that a definition of a species always enables a biologist to decide whether the forms he observes are distinct species or merely races of one species, then nobody has invented a definition. It may, moreover, be doubted that such a definition will ever be invented. Strangely enough, such a 'success' would come close to overthrowing the theory of evolution. The core of Darwin's argument was that species arise from races by a process of gradual divergence. Unless evolution had stopped [see Goudge, 1957, for a discussion of this possibility], there must be found instances of incipient species, i.e. forms that will be too distinct to be regarded races of one species but not distinct enough to be counted species. To classify them either as races or as species, is, then, arbitrary. (p. 183)

Classification in the nineteenth century

Returning to the problem of classification, during the nineteenth century classification systems became evolutionary. These, however, were not without their limitations:

(a) Though living things are classified into either plants or animals, the distinction is far from absolute. There are living organisms, in particular the Flagellates, which do not fit into either classification.

(b) There are groups which have no relationships with other groups, in respect of either structure or mode of development.

The hope that genetics could provide a definition for the term species had so far been fruitless. Here it is worth considering the views of two authorities. The first quotation is from Sturtevant, *A History of Genetics*:

One of the awkward circumstances about discussions in this field is the ambiguity in the use of the word species. There is no generally accepted definition, but most discussions have centred on the extent to which cross-sterility between populations is to be taken into account in deciding whether or not two groups are to be considered as separate species. The nature and origin of interspecific sterility has been recognised since Darwin as one of the major problems of evolution. (p. 111)

The second is from Dobzhansky's *Mankind Evolving:*

Any genetic agencies that decrease or prevent gene exchange between species are called reproductive isolating mechanisms. Many such mechanisms are known. They range from differences in preferred habitats, behaviour, courtship rituals, and breeding seasons, to difficulties of fertilisation and inviability or sterility of the hybrids. No one of these isolating mechanisms occurs universally between all species: not all species are 'intersterile'; nor do all lack sexual attraction for each other; nor do all differ in habitat preferences or in mating seasons. Any one isolating mechanism may be strong enough to isolate a species, but the separation of species is usually accomplished by a combination of several mutually reinforcing mechanisms.

Species in sexual organisms are, then, closed breeding communities, i.e. systems of populations having access to a common gene pool but reproductively isolated from other populations with separate gene pools. Biologists have been talking and writing about species for at least two centuries. The definition just given is couched in the idiom of modern biology and particularly genetics. There has been some divergence of opinion as to whether species should be defined this way. The controversy need not concern us here, since what really matters for our present purpose is the fact that there is a stage in the evolutionary divergence of populations when genetically open systems (races) become genetically closed systems (which we may call species, or else invent a different name). (pp. 184–5)

EARLY EVOLUTIONARY SPECULATIONS

Who first considered the idea of evolution? Perhaps claims could be made for many, but Aristotle and Buffon are prominent contenders. Here one can present and discuss the views of two historians of science. In *A History of Biology* Singer has this to say about Aristotle:

It is probable that in ascribing to animals certain human qualities, Aristotle was influenced by his advance towards what would nowadays be called a belief in Evolution. . . . It cannot be said that he ever definitely attained to the 'evolutionary' point of view. But it is evident that he was moving in that direction, and perhaps if he had lived another ten years he might have reached it. But without claiming him as an Evolutionist, it is quite easy to read an evolutionary meaning into some of his biological writing. To do so is to develop but not to force his meaning. (pp. 39–40)

The case for Buffon is discussed by Wilkie in *Darwin's Biological Work, Some Aspects Reconsidered,* edited by P. R. Bell:

. . . I do not think he can be called an originator or founder of the idea, and on the whole there seems to be small reason to suppose that he even subscribed to it. He seems to have regarded it as an interesting speculation of which a lot might be made some day, but which certainly could not be established upon the basis of such facts as were known to him. This seems to me to be exactly the position which a competent naturalist should have adopted during the eighteenth century, and to say that Buffon did adopt it is, I think, in no way to detract from his stature, rather, indeed, to enhance it. But the relation of Buffon to the theory of evolution is by no means merely that of a prudent judge who regarded it as non-proven. His services to the cause of the theory were not unimportant. He attempted a natural explanation of the origin of the earth in general, and of sedimentary rocks in particular, and thus did for Lamarck much what Lyell did for Darwin: he fostered the idea of a gradual temporal development to natural causes. He directed attention to the problems associated with the theory of evolution, and he was, as far as I am aware, the first to appreciate that the geographical distribution of animals contained an essential part of the evidence relating to the origin of species. Finally, he recognised that many species of animals had become extinct before the appearance of man on earth, and that there were, consequently, causes of extinction which owed nothing to human agencies. This is a not unimportant element in the case for evolution, and it is one which Lamarck failed to understand. (p. 287).

Reference can also be made to *Forerunners of Darwin, 1745–1859*.

The Lamarckian theory of evolution

There is no doubt that Lamarck was greatly influenced by Deist philosophy, as has already been suggested. This was the conception of the 'chain of beng' with its underlying principles of plenitude, continuity and gradation. This philosophy did not have its origin in the eighteenth century. Although perhaps it was more widely accepted then, its origins go back to the time of Plato and Aristotle. It is interesting to note that though the influence of Aristotelian biology had been on the decline for a long time, the doctrine of natural species was kept alive through the chain of being. In *Darwin's Biological Works* Wilkie states very briefly the thought of the Deists:

The central tenet of Deism is that the relation of God to the created world is essentially 'rational', that is, that the ways of God in dealing with his creatures are comprehensible to the unaided human reason. With this central notion were associated a number of sub-ordinate propositions which could follow logically from the basic assumption only if it were also supposed that all God's purposes were known, and that to achieve them he would use only the means which recommended themselves to the common sense of the average enlightened man of the time. (p. 267)

The chain was not without its positive influences. It did stimulate biologists to look for 'missing links' that were apparent in the chain.

Jean Baptiste de Monet Lamarck (1744–1829) gave three accounts of his theory of evolution, published in 1801, in 1809 (in *Philosophie Zoologique*) and also in 1815 in his enlarged treatise on the invertebrates. This third treatment appears unaltered in the 1835 edition. The four laws of Lamarck have been paraphased by Wilkie:

1. Nature tends to increase the size of living individuals to a predetermined limit.
2. The production of a new organ results from a new need.
3. The development reached by the organs is directly proportional to the extent to which they are used.
4. Everything acquired by the individual is transmitted to its offspring. (p. 293)

The evolutionary process for Lamarck seems to depend on a single causal factor, that of the inheritance of acquired characteristics. This, however, is very much disputed by Cannon in *The Evolution of Living Things*:

Lamarck put forward his hypothesis as four separate laws and it is only one of these, the last, that expresses a belief in the inheritance of the effects of use and disuse. But as I shall show later on, it is just this law which is unnecessary. It is superfluous for it is covered by the other laws. Further, it is clear from his writings that Lamarck did not place this idea in the forefront of his evolutionary ideas. It was the second and third laws which were to him the essence of his scheme and these, as we shall see, have nothing to do with a belief in the inheritance of acquired characteristics. Thus to refer to the inheritance of acquired characteristics as the same as Lamarckism is to deprive Lamarck of what he considered his greatest contribution to evolutionary thought. Nevertheless, this is the idea which has come down to the present day. (p. 14)

In his work Lamarck does enlarge on his first law, and makes reference to a process of increased complexity in no way guided by the effects of the environment.

Lamarck's theory was never universally accepted, and with the publication of *The Origin of Species* it underwent rapid decline. The rediscovery of Mendel's work, and the genetic investigations that followed, showed that acquired characteristics could not be transmitted to the offspring.

The Darwinian theory of evolution

It is necessary at this stage to consider Darwin's background. When his voyage on the *Beagle* began in 1831, he had no reason to think that species were immutable. In fact his whole background, and also Lyell's *Principles of Geology*, which influenced him so much, would have led him to believe otherwise. During his journey three sets of observations brought Darwin to think in terms of transformation of species.

(a) He found that in the Galapagos Islands, each island had a species of finch that differed slightly from those found on the other islands.
(b) The geographical distribution of rodents.
(c) Fossils of extinct animals corresponded to animals now living.

Having arrived at a hypothesis that species had undergone evolution, he proceeded to gather additional evidence in both botanical and zoological fields. He realised that for a theory to be acceptable it must be applicable to both plants and animals. Darwin's work is a remarkable example of an inductive approach to the study of a scientific problem. He obtained further evidence from the following sources:

(a) Domesticated animals and cultivated plants.
(b) Comparative anatomy.
(c) Comparative behaviour.
(d) Embryology.
(e) Fossils.
(f) He also investigated inter-specific sterility, and found that this was not absolute.

Darwin wrote the first account of his ideas on evolution in 1842, and enlarged on this in 1844, neither of these papers being published. In his autobiography he comments on the delay in publishing, stating that if he had rushed into print in 1838, or even in 1844, it would not have been so effective as in 1859. By that time he had amassed more evidence, and there was also a change in atmosphere; his ideas were now far more likely to be accepted.

In his investigation of domesticated animals, Darwin became aware of artificial selection. The problem was how this kind of selection operated in nature. The solution of the problem came when reading Malthus' *Essay on Population*. Darwin then formulated his theory (see 'Evolution by Natural Selection' in the British Museum publication *Evolution*).

In June 1858 Darwin received a paper from Alfred Russel Wallace (1823–1913). Wallace had been working at Ternate in the Malay archipelago, and had independently arrived at conclusions similar to those of Darwin. The problem for Darwin was resolved, and Wallace's paper, an extract from Darwin's essay of 1844, and a letter from Darwin to Asa Gray, with introductions from Lyell and Hooker, were presented to the Linnean Society of London and published in the Society's journal. In 1859, less than fourteen months later, the *Origin of Species* was published (see 'The Making of the Origin' in Eiseley).

Darwin and Wallace

It is interesting to note that both Darwin and Wallace were influenced by their observations on animal distribution, by Malthus' essay, and by Lyell's *Principles of Geology*. The claims to originality are discussed by Eiseley:

Although he once generously compared his own part in the evolutionary story as one week to Darwin's twenty years, the truth was that Wallace had contemplated variation in wild nature for many years; moreover he was fully cognisant of, and sympathetic with, the evolutionary point of view. Somewhat like Darwin earlier, he had been mulling over his observations for some ten years before he was struck by this sudden flash of insight. It must be recognised, however, that although Wallace conceived his theory independently he had actually been stimulated not alone through pursuing Darwin's *Journal of Researches* but also through direct correspondence with Darwin. Darwin had written to Wallace that he agreed heartily with an earlier (1855) paper expressing evolutionary views and that 'they had thought much alike'.

Correspondence indicates that Darwin had told Wallace he was working on 'the species problem'. He generously urged Wallace on in his own speculations, but politely declined to divulge his own theory prematurely.

There is thus clear documentation which, while establishing Wallace's own claim to originality, indicates at the same time that this very astute and perceptive young naturalist was running hot on Darwin's trail. His curiosity must have been intense and there is every indication that even if the memory of Malthus provided a spark, Wallace, like Darwin, had long pondered 'the great work of Lyell'. (pp. 291–2)

It may be true to say that Wallace was stimulated by Darwin, but perhaps more important is the stimulus by Wallace which resulted in Darwin finally publishing the *Origin of Species.* Reference can also be made to a paper by H. Lewis McKinney, 'Alfred Russel Wallace and the Discovery of Natural Selection', *Journal of the History of Medicine and Allied Sciences,* **21,** 333–57 (1966).

Darwin's theory of evolution has been summarised in the following way in the *Handbook of Evolution* (British Museum—Natural History):

1. Organisms produce a far greater number of reproductive cells than ever give rise to mature individuals.
2. The numbers of individuals in species remain more or less constant.
3. Therefore there must be a high rate of mortality.
4. The individuals in a species are not all identical, but show variation in all characters.
5. Therefore some variants will succeed better and others less well in the competition for survival, and the parents of the next generation will be naturally selected from among those members of the species that show variation in the direction of more effective adaptation to the conditions of their environment.
6. Hereditary resemblance between parent and offspring is a fact.
7. Therefore subsequent generations will maintain and improve on the degree of adaptation realised by their parents by gradual change. (pp. 10–11)

Origin of Species

When the *Origin of Species* was published, Darwin was unaware of the laws of heredity. Lack of this knowledge presented difficulties. There were strong objections to the idea of 'blending inheritance', for this would lead to dilution of any new variation. The answer to the problem came when the Mendelian theory of the gene was worked out. Genes preserve their individual identity, and remain constant unless they undergo mutation. Mutations often showed discontinuous steps, and this led to disagreement between the selectionists and the geneticists. These steps were not the gradual continuous variation the selectionists were looking for. Another problem was what happens to a structure or function before it reaches a level of perfection; and here Darwin introduced the idea of gradual development with serviceable transitions. There was also the problem that variation is haphazard, and mutations take place at random. What must be stressed, however, is that according to Darwin the course of evolution is determined neither by variations nor mutations, but by the process of selection.

MODERN THEORIES

When one considers the evidence for evolution, one can see how much extra evidence has been collected since the time of Darwin—not only in those areas he investigated himself, but in the fields of physiology, genetics and serology.

When discussing modern evidence for evolution, a most profitable topic for consideration will be Kettlewell's work on *Biston betularia*. It is an excellent example of hypothesis testing, in this case testing whether melanic forms are at an advantage in smoke-polluted industrial areas. (See H. B. D. Kettlewell, 'Selection Experiments on Industrial Melanism in the Lepidoptera', *Heredity*, **10**, 287–301 (1955), and *Heredity*, **9**, 323–42 (1955); see also 'A Survey of the frequencies of *Biston betularia* and its melanic forms in Great Britain', *Heredity*, **12**, 51–72 (1958).

Photographs of the light and dark form of the peppered moth can be seen in *The Discovery of Time* (plate 9); see also p. 220.

Modern theories of evolution are now based on mutations in the genes and chromosomes. Today we also have a number of cases that demonstrate natural selection in progress.

There are, however, dissenting views, the most important of which come from R. B. Goldschmidt. His attack has been on the classical concept of the gene as well as the process of evolution suggested by the neo-Darwinians. Goldschmidt does not consider the gene to be a discrete particle, but rather an integrated part of a biochemical continuum. He does not consider that the accumulation of micro-mutations could result in the formation of a new species as a rule. He suggests there is a re-patterning of the chromosome, with an early modification of embryonic development by the new pattern. His theory indicates that very extensive changes would therefore result, and a new species could be subjected to the process of natural selection in a few steps. It should be noted that this kind of mutation has not been demonstrated. His theory also avoids the problem of the very slow rate at which evolution would take place by the accumulation of micro-mutations.

This last example demonstrates extremely well the fact that science is not a static but a dynamic activity. Theories are generalisations of the information we have at the present time. These will certainly be modified, or even rejected, as our knowledge and understanding increase.

Social Darwinism

It has been generally assumed that Darwin discovered a social as well as a biological law. It should be stressed that this was not so. One can see, however, that such terminology as 'struggle for existence' and 'survival of the fittest' are full of political connotations. Social Darwinism can often be equated with racism. The problem is discussed in T. Dobzhansky's *Mankind Evolving* and in J. Brierley's *Biology and the Social Crisis*.

Evolution of Man

In Darwin's *On the Origin of Species by means of Natural Selection on the Preservation of Favoured Races in the Struggle for Life* (1859) —to give the full title—no mention is made of the evolution of Man. His publication relating to Man came in 1871 under the title *The Descent of Man, and Selection in Relation to Sex* (see Eiseley, ch. x, 'The Reception of the First Missing Links').

The evidences for the biological evolution of Man are no different from those of other organisms. These are discussed in chapter 7 of *Mankind Evolving*. In considering human evolution the following quotation from Medawar's *The Future of Man* can be discussed:

> . . . The danger is that by calling attention to the similarities, which are not profound, we may forget the differences between our two styles of heredity and evolution; and the differences between them are indeed profound. In their hunger for synthesis and systematisation, the evolutionary philosophers of the nineteenth century and some of their modern counterparts have missed the point: they thought that great

lessons were to be learnt from similarities between Darwinian and social evolution; but it is from their differences that all the great lessons are to be learnt. For one thing, our newer style of evolution is Lamarckian in nature. The environment cannot imprint genetical information upon us, but it can and does imprint non-genetical information which we can and do pass on. Acquired characteristics are indeed inherited. The blacksmith was under an illusion if he supposed that his habits of life could impress themselves upon the genetic make-up of his children; but there is no doubting his ability to teach his children his trade, so that they can grow up to be as stalwart and skilful as himself. It is because this newer evolution is so Lamarckian in character that we are under psychological pressure to believe the genetical evolution must be so too. But although one or two biologists are still feebly trying to graft a Lamarckian or instructive interpretation upon ordinary genetical evolution, they are not nearly so foolish or dangerous as those who have attempted to graft Darwinian or purely selective interpretations upon the newer, non-genetical, evolution of mankind. (pp. 98–9)

Further, a quotation from *Mankind Evolving:*

Culture is, however, an instrument of adaptation which is vastly more efficient than the biological process which led to its inception and advancement. It is more efficient among other things because it is more rapid—changed genes are transmitted only to the direct descendants of the individuals in whom they first appear; to replace the old genes, the carriers of the new ones must gradually outbreed and supplant the former. Changed culture may be transmitted to anybody regardless of the biological parentage, or borrowed ready-made from other peoples. In producing the genetic basis of culture, biological evolution has transcended itself—it has produced the superorganic. (p. 20)

EVOLUTION AND RELIGION

It is interesting to speculate whether the controversy would have been the same if Darwin had not chosen to compare evolution with special creation in the *Origin*. What would have happened if he had taken spontaneous generation instead? Was the controversy: 'Scientific enlightenment versus theological prejudice?' as J. Hobgood puts it in his *Religion and Science*? His chapters on 'Life', 'Nothing but Apes?' and 'Science and the Bible' are very applicable to a discussion on this subject, and so is the rest of the book. Those who wish to go deeper into this question can read some of the following: *The Nature and Destiny of Man* by R. Niebuhr;

Biology and Social Theory in the Nineteenth Century by J. C. Green; and *The Phenomenon of Man* by Pierre Teilhard de Chardin.

In this section, one might consider this quotation from Eiseley:

The rise of the new science was beginning as early as the seventeenth century to erode the foundations of this Christian *mythos* and the several evolutionary debates of the nineteenth century represent only successive steps in its hastening decline. Although we may recognise the frailties of Christian dogma and deplore the unconscionable persecution of thought which is one of the less appetising aspects of medieval history, we must also observe that in one of those strange permutations of which history yields occasional rare examples, it is the Christian world which finally gave birth in a clear articulate fashion to the experimental method of science itself. Many things undoubtedly went into that amalgam: Greek logic and philosophy, the experimental methods of craftsmen in the arts as opposed to the aristocratic thinker—all these things have been debated. But perhaps the most curious element of them all is the factor dwelt upon by Whitehead—*the sheer act of faith that the universe possessed order and could be interpreted by rational minds.* For, as Whitehead rightly observes, the philosophy of experimental science was not impressive. It began its discoveries and made use of its method in the faith, not the knowledge, that it was dealing with a rational universe controlled by a Creator who did not act upon whim nor interfere with the forces He had set in operation. The experimental method succeeded beyond men's wildest dreams but the faith that brought it into being owes something to the Christian conception of the nature of God. It is surely one of the curious paradoxes of history that science, which professionally has little to do with faith, owes its origins to an act of faith that the universe can be rationally interpreted, and that science today is sustained by that assumption. (pp. 61–2)

HAS SCIENCE ANY ETHICAL VALUES?

This is a very suitable question for discussion at this stage. Reference should be made to Glass, *Science and Ethical Values*. It is often an accepted premise that science has no ethical values—an assumption Glass does not share:

This view I wish to challenge, since it is my belief that by examining critically the nature, origins, and method of science, we may logically arrive at the conclusion that science is ineluctably involved in questions of value, inescapably committed to standards of right and wrong, and unavoidably moves in the large toward social aims.

Glass also considers the ethical problems:

> the nature of the stupendous ethical problems that will face mankind in the very near future as man begins to apply his knowledge to the control of his own reproduction and future evolution.

Science has endowed man with the power to determine the course of his own future evolution. The problem of the future will indeed be ethical. Will this power be used for good or for ill?

GLOSSARY TO SECTION III

abiogenesis: living things originating from non-living substances (spontaneous generation).

biogenesis: living things originating from parents rather than from non-living substances.

chromatids: strands resulting from chromosome duplication.

chromosome: rod-shaped bodies that are stained during the process of cell division; they carry genes.

cytology: a branch of biology which involves the study of the structure and function of cells.

cytoplasm: cell material outside the nucleus.

DNA: deoxyribonucleic acid, the substance that makes up the heredity material of the chromosomes.

Drosophila: a fly widely used in genetical research.

gene: unit of heredity.

homoeostasis: tendency to maintain constant environmental conditions within cells and organisms.

meiosis: a special kind of cell division in which the number of chromosomes within the cell after division is half that of the parent cell.

metabolism: chemical activity within a living thing.

mitosis: cell division, a parent cell producing two daughter cells which are genetically identical.

mutation: a sudden change in the chemical nature of the heredity material; provides new hereditary variation.

neobiogenesis: the repeated origination of life since life began.

Neurospora: a fungus widely used in genetical research.

nucleus: the part of the cell which contains the heredity material.

phylum: one of several major groups within the animal kingdom.

Pneumococcus: a bacterium used in genetical research.

protoplasm: living matter.

Protozoa: a group of single-celled animals.

reduction division: meiosis q.v.

RNA: ribonucleic acid.

transduction: transfer of genetical material from one bacterium to another by means of a virus.

X-ray: a form of radiation having very short wave length and high energy.

Main appendices

APPENDIX A

Some further teaching schemes

In addition to courses based on consecutive topics in this guide the following outlines are offered for courses which either cut across the present arrangement or lie on the fringe of material in it.

(1) ANCIENT AND GREEK SCIENCE

(Topics I.1 and 2; II.1 and 2.)

This could form a useful course in which are compared the practical approach of the civilisations of Babylon and Egypt with the rational attitude of the Greeks. Could be developed into a general discussion of the pure/applied dichotomy.

(2) THE SEVENTEENTH CENTURY

(Topics I.4–6; II.4 and 5, with some material from 3 and 6 for contrast.)

This period abounds in interesting figures and important discoveries as well as being an important focal point in the history of science. An interesting subject which could be included here is the organisation of science—the change from individualism to cooperative effort in the scientific societies. There are several books on the Royal Society which give good background material; one of the briefest and best accounts is the first chapter by McKie in *The Royal Society, its Origins and Founders* (Royal Society, 1960).

(3) THE IMPORTANCE AND REFINEMENT OF MEASUREMENT

(Topics I.4–7; II.5, 6 and 9.)

This would require some reading outside this guide. It would probably be suitable only for scientists. Pledge in *Science since 1500* (Harper) has some interesting tables showing the increase in accuracy of measurement.

(4) CLASSIFICATION

(Topics II.6; III.5.)

A good general topic in which examples can be drawn from almost anywhere in this guide.

(5) THE ROLE OF MATHEMATICS IN SCIENCE

(Whole of Section I; topics II.2, 7 and 9.)

Done with care, such a course might help non-scientists to overcome their fear of mathematics. In addition to the topics listed above, the application of statistics in biology should be mentioned (see Bailey, N. J. J., *Statistical Methods in Biology,* E.U.P.; and Dowdeswell, W. H., *Practical Animal Ecology,* chs. 4 and 5, Methuen paperback). Many of the books in the general section of the bibliography have chapters dealing with mathematics in science.

(6) THE SCIENTIST IN SOCIETY

Under this title could be devised many courses of different range and depth, taken from almost any period. Some possibilities are: the alchemist in the Middle Ages; the Royal Society and its influence in the seventeenth century (see especially Clark, G. N., *Science and Social Reform in the Age of Newton,* Oxford); nonconformist scientists in the eighteenth and nineteenth century (also, and related to this, the 'Lunar Society'—see Schofield, R. S., *The Lunar Society of Birmingham,* Oxford); Darwin and the nineteenth century is a well-documented subject; and books on the effect of science in contemporary society are legion, e.g., Conant, J. B., *Modern Science and Modern Man* (Columbia), Jungk, R., *Brighter than a Thousand Suns* (Penguin), and the writings of Snow, Appleton, Huxley, Medawar, Bronowski, Russell, Whitehead, Weizäcker and others. Though excluded from this guide, medicine and technology are excellent subjects through which to discuss the influence of science on society. See Bibliography (appendix F), especially the general section.

(7) EXPERIMENTS

Some pupils find interesting the development of ideas as seen through key experiments, with actual demonstrations wherever practicable. Provided the warning in the introduction is heeded, such an approach can be rewarding. The choice of experiments must rest with the teacher. There are several compilations of suitable experiments available, or the teacher can devise adequate and simplified demonstrations from contemporary accounts (e.g., the *Alembic Club Reprints* and the *Harvard Case Histories*).

An alternative approach is to provide students with the results of classical experiments, make them attempt an interpretation (with guidance!) and then compare with the original interpretation. In giving guidance, sufficient must be provided to keep interest alive, but not so much that initiative is quelled.

Courses based on experiments are more likely to succeed if the experiments are quite simple and the range of ideas covered not too wide. (Remember that experimental results do not always conform to the expectations of the teacher!)

APPENDIX B

Questions and exercises

Many ideas for questions and discussion are given in the text. The following questions may provide class material for both written work and discussion. In many cases the type of question can be applied to a different subject.

The first twenty-five questions are taken from recent papers in the History and Philosophy of Science set by the Oxford and Cambridge Joint Board.

(1) What is an epicycle? How did such geometrical devices function within Ptolemy's astronomical system? How and why were such devices avoided by Kepler and Newton

(2) Would you agree that Galileo's contribution to astronomy was destructive rather than constructive?

(3) Discuss how much truth there is in Pope's couplet:

'Nature and Nature's laws lay hid in night
God said, Let Newton be! and all was light!'

(4) 'Whoever is ignorant of *motion* is ignorant of nature.' Is Aristotle's dictum supported by the history of science?

(5) Would you have been persuaded by Galileo that the earth went round the sun? (Comment on the persuasiveness of each of Galileo's arguments.)

(6) What part has the concept of atomism played in the development of physics and chemistry?

(7) Why did Young and Fresnel support a wave theory of light?

(8) How were the various attempts to classify the elements finally satisfied? What has been the significance of the periodic law for the development of chemistry in the present century?

(9) The word atom has been defined as that which *is not divided,* and also as that which *cannot be divided.* Discuss this, having regard to the views of Dalton and those of later chemists and physicists.

(10) The use of metals for practical purposes began with bronze, was followed with iron and steel, and much later with aluminium and other light metals. How do you account for this sequence?

(11) Describe some of the phenomena which the phlogiston theory attempted to explain and contrast this explanation with that put forward

by Lavoisier. Would you say that the phlogiston theory involved the wrong sort of scientific method, or that it was just simply wrong?

(12) Distinguish between a natural and an artificial classification of animals. Why is the former strong evidence that organic evolution has taken place?

(13) How have advances in chemistry and physics assisted the biologist?

(14) Describe and discuss an example of recent work on the action of natural selection in a species. Since species can be modified, why is the concept of species still valid?

(15) Discuss the work of the Greek biologists. Why were their views, though often erroneous, held in repute for so long?

(16) Aristotle is often called the 'father of biology'. Give an account of his biological work and the influence it was to have on successive ages.

(17) Describe the opening remarks in an imaginary discussion between Aristotle, Linnaeus and Darwin on the underlying patterns behind the apparent diversity found in the natural world. You should allow each of your characters to put forward his own interpretation of nature.

(18) Discuss the relative importance of the invention and the testing of hypotheses.

(19) What is the difference between a scientific and a value judgement? Do value judgements play any part in the formation of scientific hypotheses or theories? Support your answers by *one or more* examples.

(20) Give a brief account of the reasons for the renaissance of learning that took place in western Europe in the fifteenth and sixteenth centuries. What great scientific discoveries were made during this period and what was their effect on the contemporary religious and philosophical ideas?

(21) It has been said that 'a science enables us to satisfy three main needs, the condensation of our knowledge, the prediction of the future course of nature, and the explanation of natural phenomena'. Discuss.

(22) Discuss, giving historical examples, the following statement made by T. H. Huxley in 1854: 'Science is nothing but trained and organised common sense.'

(23) Write briefly about *two* of the following: (a) the importance of scientific societies, (b) team versus individual research, (c) the role of chance in scientific discovery.

(24) 'As more scientific discoveries are made so the area of our ignorance seems to increase.' Support this idea by the use of examples of

scientific discoveries which, far from solving a particular problem, served only to add to an understanding of its complexity.

(25) What is the relationship between the rise of science and the development of technology?

(26) Discuss the importance of instrumentation in the development of scientific ideas. As well as a general survey, our account should include a more detailed study of a particular instrument.

(27) Is there any difference between scientific truth, mathematical truth and historical truth?

(28) What is the proper function of mathematics in science?

(29) Compare the attitude to knowledge in a guild with that in an academy. When is it right to withhold information (trade or State secrets) and when not? (The effects of imposed secrecy on twentieth century scientists is the main theme of Jungk's *Brighter than a Thousand Suns.)*

(30) During their journeys round the western world, some of the ancient Greek manuscripts suffered translation many times. Speculate on the effect of translation on scientific writings. How do you think poetry fares by comparison?

APPENDIX C
Some general philosophical questions

Some teachers have expressed a wish for specific guidance on philosophical discussion. Here are some general questions and observations which might be borne in mind while following other schemes. These are intended as possible patterns for philosophical discussion; they are far from exhaustive, and represent only one possible (and partial) approach. As is stated in the introduction it was felt presumptuous and unwise to be too dogmatic on philosophical material: the extent and depth of discussion depend on the teacher, the class and the subject.

(1) Science can be regarded as attempts to construct logical pictures (models) by means of which we can extend and codify our experience.

This is probably a more fruitful way to regard the activity of the scientist than that of 'discovering nature'. Scientific theories are artefacts and there is no way of knowing whether or not they correspond in fact to the 'great out there'—some independently existing 'reality'. (The writings of G. E. Moore oppose this and present the 'common sense' view. They could form the basis of a critical discussion. The 'reality' problem could be introduced by a brief consideration of Berkeley and Hume. (See Russell's *History of Western Philosophy* or E. R. Emmet's *Learning to Philosophise* (Longmans)—a very good introduction for schools.)

(2) Why do scientists always plump for the simplest theory?

If one maintains the 'discovery of reality' line then it seems that nature must be simple. Is this likely? Does not nature in fact seem to be very complicated? If, on the other hand, science is devising pictures with which to think and plan ahead (*prediction* comes in here), then the simplest picture that will serve is the obvious one to choose.

One could develop here the point that 'simplest' means not more complicated than is justified by (a) the purpose in hand—e.g. it would be foolish to insist on Schrödinger's atomic model in the elementary explanation of the kinetic theory of gases, since the billiard ball atom is entirely adequate; (b) the experience at the time—e.g. the phlogiston theory was sound as an explanation of combustion in the first half of the eighteenth century, and *had* someone put forward the modern oxygen theory it would probably have been rejected as unnecessarily complicated. Notice also that for many purposes (e.g. navigation) it is sufficient to consider the earth as stationary with the heavens revolving round it. (If we accept relativity, does it matter anyway?

(3) The aesthetic appeal of economy of hypotheses (see especially topic II.2)—the maximum number of phenomena explained by the minimum number of assumptions—is probably due to simplicity: such pictures are more fruitful because they are more easily handled.

(4) Does the scientist live by faith in a 'rational universe'? What is a rational universe? Is its rationality merely that of man, or of the creator —God? Kepler thought God was a geometer (or geometry was God— see Koestler, *Watershed,* p. 60); by now then God must also be an accomplished relativistical and quantum mechanic! There are obvious difficulties here. Perhaps it is better to follow Hume and say that science works in the hope that the future will be conformable to the past. It constructs a model which fits past experience and which can be used according to rules of logic in the hope that it will fit experience to be had in the future.

There is no guarantee that it will, but this is no assumption about what nature may or may not be like. When nature does not conform, it is more profitable to change our model than to accuse her of irrational behaviour.

(5) It sometimes occurs in science that alternative theories exist which fit the facts, or sometimes one theory fits one part of the facts and another another. Such situations create difficulties for the 'discovering nature' school but, though on grounds of simplicity we may crave their 'resolution', they are entirely reasonable and indeed to be expected when science is regarded as the construction of convenient pictures. Scientific attitudes towards the wave–particle duality provide a useful study in this problem. When Frenkel said, 'Of course, according to the thinking hitherto prevalent among us, these two alternatives absolutely exclude each other. So you can believe in particles on Mondays, Wednesdays and Fridays and in waves on Tuesdays, Thursdays and Saturdays,' —he made, albeit facetiously, a good point. The 'problem' of the wave—particle duality arose because many were unwilling or unable to discard the two distinct models of 'wave' and 'particle' inherited from the nineteenth century. (Notice, however, that it is still sometimes convenient to describe the electron as a 'particle' and at other times as a 'wave'.) The contemporary description in terms of abstract mathematical equations can be regarded as a more sophisticated 'model' where the clear mental images associated with particle and wave are no longer possible.

(6) The study of a theme in science, e.g. matter theory, seems often to indicate a confusing and contradictory 'reality' of the 'outside world'. If we regard science as the development of the mind in devising explanatory models, then the confusion and contradiction is that of the human mind—surely a more reasonable supposition. Likewise the wonderful complexity and order 'of nature' revealed in the rather formidable abstractions of modern matter theory can legitimately and more profitably be regarded as indicating the sophistication and power of human thought.

APPENDIX D

Historical background

Shortage of space prevented the inclusion of adequate historical background summaries for courses based on particular periods. The extent and depth of such matters depends very much on the course and even more on the interests and ability of the teacher. Hints are given in the text and further material will be found in several of the books in the main bibliography (appendix F), especially the general section.

Here is a list of books which will be of use in themselves and will provide ample guidance to further reading.

DAMPIER, W. C., *A History of Science,* Cambridge. A splendid encyclopaedic treatment.

McKENZIE, A. E. E., *The Major Achievements of Science,* Cambridge. For historical and philosophical background the following chapter are useful: 7 (the seventeenth century), 10 (the eighteenth century), 18 (the nineteenth century), 24 (the twentieth century).

CLARK, G. N., *The Seventeenth Century,* Oxford.

WILLEY, B., *The Seventeenth-Century Background* and *The Eighteenth-Century Background,* Penguin.

BRONOWSKI, J., and MAZLISH, B., *The Western Intellectual Tradition,* Penguin.

APPENDIX E

General background material and visual aids

MUSEUMS

A comprehensive list of facilities offered by various local museums is not possible here. However, it will be to the advantage of any teacher, especially those involved in the teaching of the history of biology, to pay a visit to the local museum. Student visits to museums can also be of considerable value in widening both their knowledge and experience. The museums listed below are all well known. Those marked with an * require special arrangements to be made for a visit.

British Museum (Natural History), London S.W.7. Of special interest here is the evolution exhibit. This was prepared to mark the centenary of the publication of the Darwin–Wallace paper on evolution.

Science Museum, London, S.W.7. The biochemical part of the history of chemistry is of special interest to biologists.

*Linnean Society, London, W.1. Though not an actual museum, here are housed the collections, library, correspondence and manuscripts of Linnaeus. The Linnean Herbarium has now been recorded on microfiches. Also housed here are many of the papers of Wallace. It was at the Linnean Society that the historic paper of Darwin and Wallace was presented.

*Wellcome Historical Medical Museum and Library, London, N.W.1. A world-famous centre covering not only medicine and pharmacy, but also related subjects such as biology and chemistry.

*Museum of the Royal College of Surgeons, London, W.C.1. Here is housed the famous anatomical collection of Hunter.

Darwin House, Down House, Kent. A museum devoted to Darwin.

National Museum of Wales, Cardiff.

The History of Science Museum, Oxford.

The Tring Museum, Hertfordshire (now part of the Natural History Museum).

PHOTOGRAPHS, PRINTS ETC

The Colin Ronan Picture Library has available a very wide selection of photographs and prints of all aspects of the history of science. The address is: 27 Jeffreys Street, London, N.W.1.

SOCIETIES

Those concerned with the teaching of the history of science may wish to become members of national and international learned societies. The publications, meetings and symposia of these societies are of value. Membership is usually by election:

British Society for the History of Science, Secretary, c/o Science Museum, London, S.W.7.

British Society for the Philosophy of Science, Secretary, c/o London School of Economics, London, W.C.2.

Other societies of special interest are the following: Linnean Society of London, Burlington House, London, W.1; Eugenics Society, 69 Eccleston Square, London, S.W.1; American Genetic Association, 1507 M. St., Washington 5, D.C., U.S.A.; Society for the Study of Evolution, Department of Botany, University of California, 405 Hilgard Avenue, Los Angeles 24, U.S.A.

FILMS

This list is divided into four parts: films suitable for each of the three sections of this guide and those of general interest.

Distributors:

B.F.I.: The British Film Institute, 81 Dean Street, London, W.1.
E.F.V.A.: Educational Foundation for Visual Aids, 33 Queen Anne Street, London, W.1.
S.S.: Sound-Services Ltd., Wilton Crescent, London, S.W.19.

Others as indicated.

Section I

Isaac Newton (S.S.)
Galileo (S.S.)
Earth and Sky (B.F.I.)

Section II

The God Within (B.F.I.)
The Perfection of Matter (B.F.I.)
The Rutherford Atom (S.S.)
Atomic Physics: (E.F.V.A. and Rank Film Library, 1 Aintree Road, Perivale, Greenford, Middlesex)

(1) 1826–95
(2) Electron, positive rays and X-rays
(3) Radioactivity and nuclear structure

(4) The neutron and accelerating machines
(5) The cyclotron and nuclear fission

The Conquest of the Atom (E.F.V.A.)
History of the Discovery of Oxygen (I.C.I., Millbank, London, S.W.1)
John Dalton and the Atom (I.C.I.)

The PSCS physics and chemistry study films (B.F.I.) might be useful in illustrating certain technical points.

Section III

William Harvey and the Circulation of the Blood (E.F.V.A.)
Heredity in Animals (E.F.V.A.)
Heredity in Man (E.F.V.A.)
Industrial Melanism in the Peppered Moth (B.F.I.)
Darwin and the Insects of Brazil (Petroleum Film Bureau, 4 Brook Street, London, W.1)
Gregor Mendel (Education & TV Films, 164 Shaftsbury Ave., London, W.C.2)
Time Is (B.F.I.)
The Perception of Life (B.F.I.)
The Discovery of the Past (B.F.I.)

General

Aristotle and the Scientific Method (S.S.)
Leonardo da Vinci (S.S.)
The Revival of Learning (S.S.)
A Light in Nature (Petroleum Film Bureau)

Most of these films have been recommended as suitable for this guide. For details the distributor's catalogues should be consulted.

APPENDIX F

Bibliography

The bibliography for each section is divided into three classes:

[1] indispensable;
[2] very useful, most of these should be readily accessible;
[3] suitable for occasional reference.

There is also a general list of books and journals.

N.B. References in the text to books in class [1] are to paperback editions where possible—this is particularly relevant to *Fabric* and *Architecture* references.

SECTION I

[1]

COHEN, J. B., *Birth of a New Physics,* Heinemann Science Study Series.

HALL, A. R., *Galileo to Newton,* Collins.

KOESTLER, A., *The Watershed,* Heinemann Science Study Series. This is the most valuable section of his larger book, *The Sleepwalkers.*

TOULMIN, S., and GOODFIELD, J., *The Fabric of the Heavens,* Hutchinson and Penguin. A readable general guide, which all members of a class might have. Fairly numerous short excerpts from the sources, but not much detail in the more modern sections.

[2]

BERRY, A., *A Short History of Astronomy,* Dover Reprint. Useful for post-Newtonian topics as well as earlier work.

BRECHT, B., *Galileo,* Methuen. This could well be used for a class reading if several copies were available.

BUTTERFIELD, H., *The Origins of Modern Science,* Bell.

COLEMAN H., *Relativity for the Layman,* Penguin.

CROMBIE, A. C., *Augustine to Galileo,* Heinemann (and Mercury paperback: 2 vols.). For detailed reference.

DREYER, J. L. E., *A History of Astronomy from Thales to Kepler,* Dover Reprint.

GAMOW, G., *Gravity, Classic and Modern Views,* Heinemann Science Study Series. For the Newtonian synthesis and tides.
— *The Birth and Death of the Sun,* Mentor Books.

HALL, A. R., *The Scientific Revolution 1500–1800*, Longmans. For detailed reference rather than general reading.

HEATH, T. L., *Aristarchus of Samos*, Oxford. From the earliest Greeks to Aristarchus.

HESSE, M. B., *Forces and Fields*, Nelson.

HOYLE, F., *Astronomy*, Macdonald.
— *Frontiers of Astronomy*, Heinemann.
— *Galaxies, Nuclei and Quasars*, Heinemann.

KIRK, G. S., and RAVEN, J. E., *The Pre-Socratic Philosophers*, Cambridge.

LAROUSSE, *Encyclopedia of Astronomy*. Many excellent illustrations.

LYTTLETON, A., *The Modern Universe*, Hodder.

MANUEL, F., *Isaac Newton, Historian*, Cambridge.

NICHOLSON, M., *Science and Imagination*, Cornell. Useful for Copernicus to Newton.

PAYNE, C., GAPOSCHKIN, C. P., *Introduction to Astronomy*, Methuen paperbacks.

SAMBURSKY, S., *The Physical World of the Greek*, Routledge paperback.

SARTON, G., *A History of Science*, Dover paperbacks. Ancient science through the Golden Age of Greece. Hellenistic science and culture in the last three centuries B.C. Very valuable both for detail and general background.

SMART, W. M., *Some Famous Stars*, Longmans. Excellent for modern stellar observations.
— *The Origin of the Earth*, Penguin.

TOULMIN, S., *The Philosophy of Science*, Hutchinson.

VAUCOULEURS, G. De., *Discovery of the Universe*, Faber. Very useful for topic I.7.

[3]

BONDI, H., *The Universe at Large*, Heinemann Science Study Series.

CLARK, G. N., *Science and Social Welfare in the Age of Newton*, Oxford.

DUHEM, P., *The Aim and Structure of Physical Theory*, Oxford.

FRANKFURT, M., *Before Philosophy*, Penguin.

GAMOW, G., *The Birth and Death of the Sun*, Mentor paperbacks.

HANSON, N. R., *Patterns of Discovery*, Cambridge.

HASKINS, C. M., *The Renaissance of the Twelfth Century,* Cambridge, Mass.

HOYLE, F., *The Nature of the Universe,* Blackwell.

KEYNES, G., *Newton the Man,* Cambridge.

KUHN, T. S., *The Copernican Revolution,* O.U.P.

MORE, L. T., *Isaac Newton: A Biography,* Dover.

NEUGEBAUER, O., *The Exact Sciences in Antiquity,* Harper Torcshb. Very detailed.

NEWTON, I., *Opticks,* Dover.

OLMSTEAD, A. T., *History of the Persian Empire,* University of Chicago Press (Phoenix paperback or clothbound). Good background to Persian Empire and its relation to Greece and Egypt.

PLATO (trans. H. D. P. Lee), *Republic,* Penguin.
— *Timaeus,* Penguin.

PLEDGE, H. T., *Science since 1500,* H.M.S.O. Useful for quick reference.

RHINE, J. B., *Of E.S.P. and Telepathy,* Faber.

SANTILLANA, G. DE, *The Crime of Galileo,* Mercury paperback.

SMART, W. M., *Foundations of Astronomy,* Longmans. A more mathematical treatment.

STEBBING, L. S., *Philosophy and the Physicists,* Dover.

TRICKER, R. A. R., *The Paths of the Planets,* Mills and Boon. Contains many valuable suggestions for experiments.

SECTION II

[1]

GAMOW, G., *The Biography of Physics,* Hutchinson.
— *The Atom and its Nucleus,* Spectrum Books.
Both these are characteristically brilliant popular treatments. They provide very good explanatory analogies for the more difficult parts of physics.

HOFFMANN, B., *The Strange Story of the Quantum,* Penguin. An excellent and humorous attempt to explain the intricacies of modern physics.

PARTINGTON, J. R., *A Short History of Chemistry,* Macmillan. A very useful compendium of facts.

SAMBURSKY, S., *The Physical World of the Greeks,* Routledge paperback. Probably the best book on the subject: readable, informative and contains many useful quotations. The discussion is excellent.

STILLMAN, J. M., *The Story of Alchemy and Early Chemistry*, Dover paperback. A very full account of the history of chemistry up to about 1700. Well written and contains a large number of original quotations.

TOULMIN, S., and GOODFIELD, J., *The Architecture of Matter* (vol. 2 in 'Ancestry of Science' series), Hutchinson and Penguin. A characteristically lucid and broad treatment of matter theory. Admirable discussion and many quotations. Much of Section II is based on this text.

[2]

ALEMBIC CLUB REPRINTS, *Livingstone*. A valuable collection of original chemical papers; most of the chemical 'classics' are included.

ANDRADE, E. N. DA C., *Rutherford and the Nature of the Atom*, Heinemann, Science Study Series. A lucid and readable treatment: can be recommended as a class text.

BOHR, N., *Atomic Theory and the Description of Nature*, Cambridge paperback. A collection of essays. It is always stimulating to read the writings of the 'originals'.

DE BROGLIE, L., *The Revolution in Physics*, Routledge. A very readable and non-mathematical account of quantum and wave mechanics.

— *Physics and Microphysics*, Hutchinson.

CHILDE, V. G., *What Happened in History*, Penguin. A detailed study of the rise of early civilisations.

CONANT, J. B., *On Understanding Science*, Mentor Books. A good background book, also contains studies in pneumatics, electricity and combustion as case histories.

EDGE, D. (Ed.), *Quanta and Reality*, Hutchinson. A collection of broadcast talks by several distinguished thinkers; contains a stimulating discussion between Bohm and Pryce on quantum theory.

FARRINGTON, B., *Greek Science*, Penguin. A sound history; discussion is controversial in places.

FRANKFORT, H., *Before Philosophy*, Penguin. Concerned chiefly with early myths and speculative philosophy.

GILLISPIE, C. C., *The Edge of Objectivity*, Oxford. A collection of informative, well-written and wide-ranging essays; several are referred to in the text.

GREENAWAY, F., *John Dalton and the Atom*, Heinemann. A good biography with valuable discussion.

HARRÉ, R., *Matter and Method*, Macmillan. A detailed and rather difficult discussion of the seventeenth century; a good source of material for philosophical discussion.

HOLMYARD, E., *Alchemy,* Penguin.

LEICESTER, H. M., and KLICKSTEIN, H. S., *Source Book in Chemistry, 1400–1900,* McGraw-Hill.

LUCRETIUS, *De Rerum Natura,* Penguin.

McKIE, D., *Antoine Lavoisier, Scientist, Economist and Social Reformer,* Constable. The best biography. More on Lavoisier's chemical work will be found in the 1935 edition, but this is out of print.

PACHTER, H. M., *Paracelsus: Magic into Science,* Collier Books. A good biography; also forms a useful background to the period.

PLATO (trans. H. D. P. Lee), *Timaeus,* Penguin.

ROMER, A., *The Restless Atom,* Heinemann. Science Study Series. A readable and concise account of some twentieth-century developments.

SAMBURSKY, S., *The Physics of the Stoics,* Routledge.
— *The Physical World of Late Antiquity,* Routledge.

SHERWOOD-TAYLOR, F., *The Alchemists,* Heinemann. Well written and contains many original quotations.

TRICKER, R. A. R., *Early Electrodynamics,* Pergamon.

— *The Contributions of Faraday and Maxwell to Electrical Science,* Pergamon.
Both contain original extracts and valuable commentary.

UBBELOHDE, A. R., *Man and Energy,* Penguin. A good general survey.

WOLF, A. (new edition: McKie), *A History of Science, Technology and Philosophy in the Seventeenth & Eighteenth Centuries* (4 vols.), Allen & Unwin. A most useful source of information.

WRIGHT, S. (Ed.), *Classical Scientific Papers (Physics),* Mills and Boon. A valuable collection of many important papers of the last seventy years on atomic physics with brief and helpful comment. A bargain at two guineas.

[3]

ABERCROMBIE, M., *The Anatomy of Judgement,* Hutchinson. A good analysis of sensory perception: should be compulsory reading for all scientists.

BIRKS, J. B. (Ed.), *Rutherford at Manchester,* Heywood. An interesting set of commemorative essays and copies of several important papers.

BORN, M., *Physics in My Generation,* Pergamon. A good first-hand account of twentieth-century developments.

BOYLE, R., *The Sceptical Chymist,* Dent (Everyman).

BROCK, W. H. (Ed.), *The Atomic Debates,* Leicester.

CHAUCER, G., *Canon Yeoman's Tale,* any edition of *Canterbury Tales.* A very useful account of contemporary alchemy; might be used as a text for topic II.3.

CLERK MAXWELL, J., *Electricity and Magnetism,* Oxford. It is often hard work but always worthwhile reading an original.

CONANT, J. B., *Harvard Case Histories* (2 vols.), Harvard. A collection of detailed studies of discoveries. Any one of these might form the basis for a course.

CROSLAND, M., *Historical Studies in the Language of Chemistry,* Heinemann. A detailed work with much useful information.

CURIE, E., *Madame Curie,* Heinemann.

DALTON, J., *A New System of Chemical Philosophy,* Oldbourne or Citadel.

FARADAY, M., *Experimental Researches in Electricity,* various complete versions, extracts in Dent (Everyman).

GIBBS, F. W., *Joseph Priestley,* Nelson. A general biography.

DE GIVRY, G., *A Pictorial Anthology of Witchcraft, Magic and Alchemy,* Spottiswoode (University Press, New York). Many magnificent illustrations.

GOMPERZ, T., *The Greek Thinkers* (4 vols.), Murray paperback. A very comprehensive survey.

GRANT, M. (Ed.), *Birth of Western Civilisation,* Thames & Hudson. Contains a great many useful illustrations.

GUERLAC, H., *Lavoisier: the Crucial Year,* Cornell.

HALES, S., *Vegetable Staticks,* Oldbourne.

HANSON, N. R., *Patterns of Discovery,* Cambridge.
— *The Concept of the Positron,* Cambridge.
Two sophisticated works in the philosophy of science; there is some valuable material for discussion.

HEISENBERG, W., *Philosophical Problems of Nuclear Physics,* Faber.
— *The Physicist's Conception of Nature,* Hutchinson.
Both are readable and stimulating.

HESSE, M., *Forces and Fields,* Nelson. A comprehensive and sophisticated survey.

HOOKE, R., *Micrographia.* Dover paperback (in fascimile).

KNIGHT, D. M., *Atoms and Elements,* Hutchinson. A critical re-examination of matter theory in the nineteenth century.

LAVOISIER, A., *The Elements of Chemistry,* Dover paperback. Contains an admirable introduction.

MEDAWAR, P. B., *The Art of the Soluble,* Methuen. A collection of splendidly controversial essays.
— Article 'Is the Scientific Paper a fraud?' published in *Experiment,* B.B.C.

MILLIKAN, R., *The Electron,* Phoenix paperback. A readable account of its discovery and properties.

NEWTON, I., *Opticks,* Dover paperback.

NORTON, J., *The Ordinall of Alchemy,* Arnold (in facsimile). An interesting work on later medieval alchemy.

PIGGOTT, S. (Ed.), *Dawn of Civilisation,* Thames & Hudson. Admirably illustrated.

POTTER (publisher), *On Modern Physics.* A collection of essays by Heisenberg, Born, Schrödinger and Auger.

WHITE, J. H., *The Phlogiston Theory,* Arnold. A fairly detailed and well reasoned account of the rise and fall of phlogiston.

WILLIAMS, L. Pearce, *Michael Faraday,* Chapman & Hall. A most important study of Faraday's significance in science.

WOOLLEY, L., *Art of the World: Mesopotamia and the Middle East.*
METHUEN, An interesting text and some superb illustrations for topic II.1.

SECTION III

[1]

BELL, P. R. (Ed.), *Darwin's Biological Work,* Cambridge. Essential for the historical assessment of evolution.

CANNON, H. G., *The Evolution of Living Things,* Manchester University Press. The views of a modern Lamarckist, interesting and stimulating.

CARLSON, E. A., *The Gene: A Critical History,* Saunders. This is probably one of the most important books to be published in the history of genetics. Carlson has the advantage of having worked under Muller. He writes from the point of view of the history and philosophy of science. Much of this book will need considerable knowledge of genetics.

DOBZHANSKY, T., *Mankind Evolving,* Yale. Deals with such topics as evolution of Man, human genetics, social evolution etc. Excellent and very readable.

EISELEY, L., *Darwin's Century,* Gollancz. An excellent book with a good account of Darwin's and Wallace's work. There is also a good historical account of the background to Darwinism.

HUGHES, A., *A History of Cytology,* Abelard-Schuman. No serious study of the history of cytology is possible without this book.

KEOSIAN, J., *The Origin of Life,* Chapman & Hall. He uses the historical approach to trace the nature of the problem, while at the same time making an evaluation in terms of modern research.

PETERS, J. S. (Ed.), *Classic Papers in Genetics,* Prentice-Hall. An important collection of papers, bringing together reprints of papers by Mendel, Johannsen, Bateson, Morgan, Sturtevant, Muller and others.

SINGER, C., *A History of Biology,* Abelard-Schuman. Takes the history of biology up to 1900; impossible to be without this book.

STURTEVANT, A. H., *A History of Genetics,* Harper & Row. Sturtevant has himself made significant contributions to the development of genetics. The book goes up to the late 1940s. He treats the subject largely from a biographical point of view.

TOULMIN, S., and GOODFIELD, J., *The Architecture of Matter,* Hutchinson & Penguin. Part III of this book, 'The Structure of Life', is concerned with the history of biology.

— *The Discovery of Time,* Hutchinson and Penguin. Evolution is a major theme in this book. It starts from the classical period, and moves into eighteenth-century geology and nineteenth-century zoology.

[2]

ASIMOV, L., *A Short History of Biology,* Nelson. A short but readable account of biological ideas up to the present day.

— *The Genetic Code,* Murray. The story of DNA, the chain of life.

BRIERLEY, J., *Biology and the Social Crisis,* Heinemann. Deals with modern world problems, populations, genetics, race, etc.

DUNN, L. C., *A Short History of Genetics,* McGraw-Hill. Traces the development of the main genetical developments during the period 1864–1939.

GEIKIE, A., *The Founders of Geology,* Dover. One of the best accounts of the history of geology, it also includes accounts of the influences of geology on the development of biology.

GLASS, B., *Science and Ethical Values,* Oxford. An excellent discussion.

HABGOOD, J., *Religion and Science,* Mills & Boon. A well balanced introduction to the subject.

HUXLEY, J., *The Living Thoughts of Darwin,* Cassell. Introduces some of the most important passages from the *Origin* and *Descent of Man* ; extremely useful.

KELLY, P., *Evolution and its Implications,* Mills & Boon. A very readable book; traces the development of genetics and gives an account of the evidence for evolution: particular attention to Man.

MEDAWAR, P. B., *The Future of Man,* Methuen. BBC Reith lectures; a stimulating book.

PLEDGE, H. T., *Science since 1500,* H.M.S.O. An extremely interesting and useful source of reference; covers the development up to recent times of many biological topics.

SINGER, C., *A Short History of Scientific Ideas up to 1900,* Oxford. An extremely useful source of reference.

—*A Short History of Anatomy and Physiology from the Greeks to Harvey,* Dover. An excellent reference for this period.

[3]

BROCK, T., *Milestones in Microbiology,* Prentice-Hall. Includes papers on spontaneous generation by Leeuwenhoek, Needham, Spallanzani, Pasteur and Lister.

CAIN, A. J., *Classification of Animals,* Hutchinson.

CONANT, J. B., *Harvard Case Histories in Experimental Science,* Harvard. Deals with two important topics in the history of biology: Pasteur's study of fermentation, and Pasteur's and Tyndall's study of spontaneous generation.

DARWIN, C., *On the Origin of Species,* Harvard. A facsimile of the first edition of the *Origin,* with introduction.

FISHER, R. A., *Experiments in Plant Hybridisation,* Oliver & Boyd. Contains Mendel's paper and Fisher's papers on Mendel.

GABRIEL, M. L., and FOGEL, S., *Great Experiments in Biology,* Prentice-Hall. Includes papers on: cell-theory (Hooke, Brown, Schwann); microbiology (Leeuwenhoek, Pasteur, Koch); embryology (Redi, Spallanzani, Weismann, Driesch); genetics and evolution (Mendel, Morgan, Muller, Beadle, Darwin, Wallace).

GOODFIELD, J., *The Growth of Scientific Physiology,* Hutchinson.

HARVEY, W., *Circulation of the Blood,* Dent (Everyman). This is considered to be one of the best translations.

OPARIN, A. I., *Origin of Life,* Dover. Morgulis' translation with up-to-date evaluation in the light of modern research.

TAYLOR, A. E., *Aristotle,* Dover.

GENERAL BIBLIOGRAPHY

These books will provide useful background reading for any course based on this guide.

BECK, S. D., *The Simplicity of Science,* Lutterworth Press and Penguin. A general survey, very basic; begins 'What is experiment?' 'What is theory?' Contains useful material for discussion.

BEVERIDGE, W. I. B., *The Art of Scientific Investigation,* Mercury Books. An analysis of the anatomy of scientific discovery. Concerned chiefly with biological problems.

BRIERLEY, J. (Ed.), *Science in its Context,* Heinemann. A general guide to science as a liberal study; contains many useful and stimulating articles.

BURNISTON-BROWN, G., *Science, its Method and its Philosophy,* Allen & Unwin. A simple and interesting introduction to the philosophy of science.

DAVIES, J. T., *The Scientific Approach,* Academic Press.

GREEN, M., *Science and the Shabby Curate of Poetry,* Longmans. Stimulating, though disorganised, essays on the two cultures.

HALL, A. R., *The Scientific Revolution,* Longmans. An excellent general account of the period 1500–1800.

HARRÉ, R., *An Introduction to the Logic of the Sciences,* Macmillan. A readable and lucid introduction.

— *The Sciences: Their Origin and Methods*, Ed., Blackie.

HARRISON, J., *Scientists as Writers,* Methuen. A collection of extracts from the writings of a wide variety of scientists conveniently grouped under eleven headings. Excellent material for class use. The author provides many good points for discussion and an extensive annotated bibliography.

LOVEJOY, A. O., *The Great Chain of Being,* Harper Torchbooks. A deep, detailed yet readable account of man's attempts to understand the rational nature of things.

MICHEL, H., *Scientific Instruments in Art and History,* Barrie & Rockliff. A full survey, lavishly and beautifully illustrated.

NAGEL, E., *The Structure of Science,* Routledge. A deep and detailed analysis, yet for the most part eminently readable; could provide useful ideas for more sophisticated discussions.

PLEDGE, H. T., *Science since 1500,* Harper Torchbooks. An adequate conventional treatment with some admirable unconventional 'cross current' discussions.

POPPER, K., *The Logic of Scientific Discovery,* Hutchinson. A classic in the philosophy of science. Admirable for sophisticated discussions.

PRICE, D. J. DA S., *Big Science : Little Science,* Columbia.

PRICE, D. K., *The Scientific Estate,* Harvard.

SAVORY, T. H., *The Language of Science,* André Deutsch. A fascinating study of the origins and development of scientific terms. Would form the basis of a useful 'fringe' topic.

SHERWOOD-TAYLOR, F., *Science Past and Present,* Mercury Books. A good and well-written general survey.

STONE, K., *Evidence in Science,* J. Wright and Sons. Designed as a 'primer' for medical students, but its accuracy and precision will recommend it to a much wider audience.

TATON, R., *Reason and Chance in Scientific Discovery,* Hutchinson. An excellent analysis of the anatomy of discovery. It contains an abundance of illustrative material and should be of particular value in discussing the 'linking themes' mentioned in the introduction.

TAYLOR, A. M., *Imagination and the Growth of Science,* Murray. A successful attempt to dispel the 'too tidy' view of scientific investigation. It deals chiefly with contemporary matter theory. Has a useful collection of biographies.

TOULMIN, S., *The Philosophy of Science,* Hutchinson.

TRICKER, R. A. R., *The Assessment of Scientific Speculation,* Mills & Boon. Readable and contains much interesting comment.

(A few of the books in this bibliography may not be in print but are available in public libraries)

Publications

Listed below are some important journals in the English language:

Isis
Annals of Science
History of Science
British Journal for History of Science
British Journal for the Philosophy of Science
Ambix
Chymia
Journal of History of Biology (Harvard)
Journal of History of Ideas
Journal of History of Medicine and Allied Science

Occasionally interesting papers appear in:

Endeavour
Scientific American
New Scientist
Journal of Chemical Education
School Science Review
Bulletin of General Studies Association

Pamphlets suitable for student reading:

B.S.C.S. pamphlets (Harrap)
Scientific American offprints (Freeman)

INDEX